INTEGRAL METHODS IN NONLINEAR DYNAMICS OF SYSTEMS

Series on Advances in Mathematics for Applied Sciences – Vol. 96

INTEGRAL METHODS IN NONLINEAR DYNAMICS OF SYSTEMS

A. A. Martynyuk

S.P. Timoshenko Institute of Mechanics,
National Academy of Sciences of Ukraine, Ukraine

World Scientific

NEW JERSEY · LONDON · SINGAPORE · BEIJING · SHANGHAI · HONG KONG · TAIPEI · CHENNAI · TOKYO

Published by

World Scientific Publishing Co. Pte. Ltd.

5 Toh Tuck Link, Singapore 596224

USA office: 27 Warren Street, Suite 401-402, Hackensack, NJ 07601

UK office: 57 Shelton Street, Covent Garden, London WC2H 9HE

Library of Congress Control Number: 2025031766

British Library Cataloguing-in-Publication Data
A catalogue record for this book is available from the British Library.

Series on Advances in Mathematics for Applied Sciences — Vol. 96
INTEGRAL METHODS IN NONLINEAR DYNAMICS OF SYSTEMS

Copyright © 2026 by World Scientific Publishing Co. Pte. Ltd.

ISBN 978-981-98-1799-3 (hardcover)
ISBN 978-981-98-1800-6 (ebook for institutions)
ISBN 978-981-98-1801-3 (ebook for individuals)

For any available supplementary material, please visit
https://www.worldscientific.com/worldscibooks/10.1142/14442#t=suppl

Desk Editors: Murali Appadurai/Srinidhi Murugan

Typeset by Stallion Press
Email: enquiries@stallionpress.com

Always Remember and Love:
In Memory and Dedication to the Wife, Mother, and Grandmother
Alevtina
(1942–2024)

Preface

It is known that mathematical modelling of real processes in mechanical and other systems leads to the consideration of linear or nonlinear systems of ordinary differential equations or equations with partial derivatives. The properties of solutions to such systems are investigated using appropriate methods developed or adapted for this class of equations. One of the methods for analysing the solutions to systems of differential equations is the integral method. This method is based on linear or nonlinear integral inequalities and the representation of solutions in integral form based on the Cauchy matrix, followed by the application of the corresponding integral inequality. At the same time, the Gronwall–Bellman lemma and its generalisations for the linear integral inequality are most widely used.

This monograph presents an integral method for analysing the dynamic behaviour of nonlinear non-stationary and controlled systems. The method is based on the application of nonlinear integral inequalities in obtaining new estimates for the norms of solutions and Lyapunov functions on solutions of the corresponding systems of differential equations of perturbed motion.

The monograph consists of a preface, seven chapters, two appendices, a bibliography, and a subject index.

In the first chapter, new boundaries are established for solutions to systems of ordinary differential equations, infinite systems of differential equations, and also for ordinary differential equations with a small parameter on the right-hand side of the equations. The general approach is based on the application of nonlinear integral inequalities and their representation in a pseudo-linear form.

In the second chapter, a system of equations of perturbed motion of a general form is considered, along the solutions of which new estimates for the Lyapunov function are established and new conditions for various types of boundedness of motion are indicated. For two coupled systems of equations, the conditions for boundedness of motion with respect to some variables are established. Conditions for the practical boundedness of motion with respect to given areas of initial and subsequent disturbances are also obtained.

The third chapter is devoted to the dynamic analysis of the motion of polynomial systems. Namely, estimates of the variation of Lyapunov functions for polynomial systems are established, on the basis of which sufficient conditions for various types of motion stability are formulated. Here, the conditions for stabilisation of motion to various types of stable motion are also obtained. The chapter ends with an analysis of the stability of the zero solution of a polynomial system with aftereffect.

The fourth chapter proposes a constructive application of nonlinear integral inequalities in estimating the boundaries of motion of nonlinear systems under interval initial conditions. Here, we also consider the problem of stabilising the motion of a system with many control elements.

The fifth chapter considers the class of quasi-linear systems with a fractional-like derivative of the system's state vector. For this class of fractional-like equations (FLEs), estimates of the Lyapunov function on the solutions of the equations under consideration are established, and sufficient conditions for boundedness and stability of motion according to Lagrange are indicated.

The sixth chapter discusses an integral method for dynamic equations with a fractional-like derivative of the state vector on the time scale. Here, new estimates are established for the variation of the Lyapunov functions along the solutions of FLEs based on integral inequalities on the time scale. The estimates obtained are used to analyse various types of stability and boundedness of solutions of dynamic equations with a fractional-like derivative.

The seventh chapter examines the problems of equilibrium stability in a generalised model of confrontation between two countries and alliances with an expanded content of functions describing hostility between countries. The studies are conducted using the direct Lyapunov method associated with nonlinear integral inequalities. Based on the estimates of the norm of the armament vector, the conditions for equilibrium stability are indicated, as well as the conditions for an increase (decrease) in the level of armament of the countries involved in the confrontation.

Appendix A provides some information from interval analysis necessary to present the results in Chapter 4.

Appendix B gives an efficient estimate of the constant in the exponential function, which is used in many problems of the qualitative analysis of quasi-linear systems.

This book presents the following for the first time:

(1) A technique for estimating the norms of solutions to systems of equations of perturbed motion of some classes of nonlinear non-stationary systems of equations has been developed based on the constructive application of nonlinear integral inequalities.
(2) New sufficient conditions for various types of stability and boundedness of motion of the considered systems of equations of perturbed motion are obtained.
(3) A generalisation of the direct Lyapunov method is obtained based on new estimates for the variation of the Lyapunov function for the dynamic analysis of solutions of various types of systems of nonlinear equations.
(4) A technique is developed for analysing the robust stability of motion for non-autonomous affine systems of differential equations with many control elements under interval initial conditions.
(5) New conditions for the stability of the equilibrium state in the generalised Richardson model of a confrontation between two countries and alliances are obtained.

This book is intended for specialists working in the field of qualitative analysis of systems of differential equations. The application of the general results presented in the book is related to the necessity of constructing suitable estimates for the Lyapunov functions on solutions of systems of equations or using estimates for the norms of the solutions of the equations under consideration.

About the Author

Anatoliy A. Martynyuk is a renowned specialist in the field of stability theories and nonlinear mechanics. In 1967, he defended his PhD thesis, and in 1973, his doctorate in physical and mathematical sciences at the Institute of Mathematics of the National Academy of Sciences of Ukraine. In 1977, he created the Department of Process Stability at the S.P. Timoshenko Institute of Mechanics of the National Academy of Sciences of Ukraine and has been its head ever since. In 1992, he initiated the creation of the international series of scientific monographs *Stability and Control: Theory, Methods and Applications* (Great Britain). The series has published 22 volumes, which have received worldwide recognition. In 2001, Dr. Martynyuk founded the international academic journal *Nonlinear Dynamics and Systems Theory* (with its online version at www.e-ndst.kiev.ua) and is its editor-in-chief. In 2006, he founded a new international series of scientific monographs, textbooks, and lecture courses entitled *Stability, Oscillations and Optimization of Systems*, published by Cambridge Scientific Publishers (UK), and is its editor-in-chief. To date, 11 volumes of the series have been published. He is the author (or co-author) of several monographs and books in English, Chinese, and Russian, devoted to the problems of stability and control of nonlinear dynamic systems. In 2008, Dr. Martynyuk was awarded the State Prize of Ukraine in the field of Science and Technology, and in 2009, he was elected as a full member of the National Academy of Sciences of Ukraine. According to the 2024 ScholarGPS Top Scholar distinction, he is among the top 0.5% of scientists globally in nonlinear systems.

Acknowledgements

We are immensely grateful to Professors A. Yu. Alexandrov, S. V. Babenko, M. Bohner, A. G. Mazko, D. D. Šiljak, and I. Stamova, who have read separate sections of the book and made a number of important comments and suggestions.

The staff of the Department of Process Stability at the S. P. Timoshenko Institute of Mechanics of the National Academy of Sciences of Ukraine, E. G. Golub, T. O. Lukyanova, and S. N. Rasshyvalova, made considerable effort in preparing the manuscript for publication.

The author sincerely thanks the specialists mentioned above as well as the editors and production department of the publishing house for their interest in the project.

A. A. Martynyuk
Kyiv, Ukraine, 2025

Contents

Chapter 1

Novel Boundaries for Solutions of Nonlinear Systems with Applications

1.1 Introduction

In this chapter, new boundaries are established for solutions to systems of ordinary differential equations, infinite systems of differential equations, and also for ordinary differential equations with a small parameter on the right-hand side of the equations. The general approach is based on the application of nonlinear integral inequalities.

Section 1.2 considers a bilinear system of differential equations and establishes conditions for β-boundedness and (ε, δ, J)-stability of unperturbed motion.

Section 1.3 investigates quasi-linear non-autonomous systems of ordinary differential equations and studies the qualitative properties of unperturbed motion.

In Section 1.4, sufficient conditions for boundedness of motion with respect to some variables are obtained. The general results are used in the study of systems of equations with fast and slow variables.

Section 1.5 considers the initial problem for an infinite system of ordinary differential equations, including systems in the standard Bogolyubov form. Here, estimates for the norm of solutions are obtained, and the problem of the proximity of solutions to the original and averaged systems of equations is studied.

The final section provides comments and bibliography for this chapter.

1.2   Qualitative Analysis of Solutions to Bilinear Systems

The basis of the integral method of qualitative analysis of solutions of linear and quasi-linear systems of equations of perturbed motion are linear integral inequalities. Let us recall some of them.

In 1919, *Gronwall* [47] obtained the following result.

Lemma 1.1. *Let a real continuous function $x(t)$ on the interval $J = [\alpha, \alpha + h]$ satisfy the inequality*

$$0 \le x(t) \le \int_{\alpha}^{t} (a + bx(s))ds \qquad (1.1)$$

for all $t \in J$, where a and b are non-negative constants. Then,

$$x(t) \le ah \exp(bh) \qquad (1.2)$$

for all $t \in J$.

This result initiated a large number of new integral inequalities (see [14, 18, 144] and the bibliography therein) in this direction.

Another inequality similar to that in Lemma 1.1 was established in 1943 by *Bellman* [14].

Lemma 1.2. *Let the functions $x(t)$ and $k(t)$ be continuous and non-negative for all $t \ge \alpha$, and let α be a non-negative constant. Then, from the inequality*

$$x(t) \le a + \int_{\alpha}^{t} k(s)x(s)ds, \quad t \ge \alpha, \qquad (1.3)$$

it follows that

$$x(t) \le a \exp \left(\int_{\alpha}^{t} k(s)ds \right), \quad t \ge \alpha. \qquad (1.4)$$

Since $\int_{\alpha}^{t} ads \le ah$ for $\alpha \le t \le \alpha + h$, it is clear that the Bellman bound includes Gronwall's inequality.

Note that perhaps the first work in which the integral inequality was used in the study of differential equations was the paper [135]. This paper presents some general results for the theory of differential equations; in particular, the concepts of the maximum and minimum solutions of a system of differential equations were introduced.

The monograph [15] presents the main results of the analysis of boundedness, asymptotic behaviour, and stability of solutions of non-autonomous linear and quasi-linear systems based on Lemma 1.2.

When studying the properties of solutions of nonlinear systems, the following lemma is used (cf. [2]).

Lemma 1.3. *Let the functions $x(t)$ and $k(t)$ be continuous and non-negative for all $t \geq \alpha$, and let a be a non-negative constant $p > 1$. Then, from the inequality*

$$x(t) \leq a + \int_{\alpha}^{t} k(s) x^{p}(s) ds, \quad t \geq \alpha, \tag{1.5}$$

it follows that

$$x(t) \leq a(1 - N(\alpha, t))^{-\frac{1}{p-1}} \tag{1.6}$$

for all $t \geq \alpha$, for which

$$N(\alpha, t) = (p-1)a^{p-1} \int_{\alpha}^{t} k(s) ds < 1.$$

The original proof of Lemma 1.3 is given in the paper by N'Doye [124].

1.2.1   Statement of the problem

The following *bilinear system* of equations of perturbed motion is considered:

$$\frac{dx}{dt} = A(t)x + B(t, x)x, \tag{1.7}$$

$$x(t_0) = x_0, \tag{1.8}$$

where $x \in \mathbb{R}^n$, $A(t)$ is an $n \times n$ matrix with continuous elements, $B(t, x)$ is an $n \times n$ matrix with elements given for all $(t, x) \in \mathbb{R} \times D$, and $D \subseteq \mathbb{R}^n$ is an open set containing the state $x = 0$.

Here and further in the text, the Euclidean norm of the vector x and the spectral norm of the matrix are used.

Consider the system (1.7) under the following assumptions:

A_1. There is a non-negative integrable function $a_1(t)$ for all $t \geq t_0 \geq 0$ such that

$$\|A(t)\| \leq a_1(t) \quad \text{for all } t \geq t_0 \geq 0.$$

A_2. There is a non-negative integrable function $a_2(t)$ for all $t \geq t_0 \geq 0$ such that $\mathbb{R}_+ \times D$ has the estimate

$$\|B(t, x)\| \leq a_2(t)\|x\|.$$

Under the conditions from assumptions A_1 and A_2, it is of interest to obtain an *estimate* for the *norm of solutions* to the initial problem (1.7)–(1.8).

1.2.2   Lemma on estimating the norm of solutions

The following assertion holds.

Lemma 1.4. *Let the conditions for the existence of a solution* $x(t) = x(t, t_0, x_0)$ *of the initial problem (1.7)–(1.8) be satisfied and conditions A_1 and A_2 hold. Then, the solution norm satisfies the estimate*

$$\|x(t)\| \leq \|x_0\| \exp\left(\int_{t_0}^{t} a_1(s)ds\right)(1 - \Phi(t_0, t))^{-1} \qquad (1.9)$$

for all $t \in J$, for which

$$\Phi(t_0, t) = \|x_0\| \int_{t_0}^{t} a_2(s) \exp\left(\int_{t_0}^{s} a_1(\tau)d\tau\right) ds < 1. \qquad (1.10)$$

Proof.   Rewrite the system (1.7) in the integral form:

$$x(t) = x_0 + \int_{t_0}^{t} (A(s) + B(s, x(s)))\, x(s)ds. \qquad (1.11)$$

In accordance with conditions A_1 and A_2, we rewrite the relation (1.11) in the form

$$\|x(t)\| \leq \|x_0\| + \int_{t_0}^{t} (a_1(s) + a_2(s)\|x(s)\|)\, \|x(s)\|ds. \qquad (1.12)$$

$\square$

Applying Lemma 1.2 to the inequality (1.12), we obtain the estimate

$$\|x(t)\| \leq \|x_0\| \exp\left(\int_{t_0}^{t} a_1(s) + a_2(s)\|x(s)\|\right) ds. \qquad (1.13)$$

We transform the estimate (1.13) to the form

$$-\|x(t)\| \exp\left(\int_{t_0}^{t}(-a_2(s)\|x(s)\|)ds\right) \geq -\|x_0\| \exp\left(\int_{t_0}^{t} a_1(s)ds\right). \qquad (1.14)$$

Next, we multiply both sides of the inequality (1.14) by $a_2(t)$ to get

$$\frac{d}{dt}\left[\exp\left(-\int_{t_0}^{t}(-a_2(s)\|x(s)\|)ds\right)\right] \geq -\|x_0\|a_2(t)\exp\left(\int_{t_0}^{t} a_1(s)ds\right). \qquad (1.15)$$

Integrating the inequality (1.15) between t_0 and t and setting $e^{t_0} = 1$, we get

$$(\|x(t)\|)^{-1} \|x_0\| \exp\left(\int_{t_0}^{t} a_1(s)ds\right)$$

$$\geq 1 - \|x_0\| \int_{t_0}^{t} a_2(s) \geq \exp\left(\int_{t_0}^{s} a_1(\tau)d\tau\right) ds. \qquad (1.16)$$

The inequality (1.16) implies the estimate (1.9) under the condition (1.10). Thus, Lemma 1.4 is proved.

1.2.3 Boundedness and stability of motion

The estimate (1.9) allows one to indicate sufficient conditions for the type of boundedness defined in the following for all solutions of the system (1.7) under the initial conditions (1.8). Assume that $B(t,x) \neq 0$ for $x = 0$.

Definition 1.1. The solutions $x(t)$ of the system (1.7) are *β-bounded* if, for a given $\beta > 0$ and $0 \leq \|x_0\| < \infty$, we have the estimate

$$\|x(t)\| < \beta \quad \text{for all} \quad t \in J,$$

where β may depend on each solution of the system (1.7).

The classical definition of *boundedness of solutions* (see [169]) only assumes the existence of the constant $\beta > 0$, which appears in Definition 1.1.

The following statement holds true.

Theorem 1.1. *Let, in the system* (1.7), $B(t,0) \neq 0$, *the conditions of assumptions* A_1 *and* A_2 *be satisfied, and the inequality* (1.10) *hold for all* $t \in J$.
Then, if

$$\exp\left(\int_{t_0}^{t} a_1(s)ds\right) (1 - \Phi(t_0,t))^{-1} < \frac{\beta}{\|x_0\|} \qquad (1.17)$$

for all $x_0 \in \mathbb{R}^n$, $\|x_0\| < \infty$, *then all solutions* $x(t,t_0,x_0)$ *of the system* (1.7) *are β-bounded.*

Proof. Under conditions A_1 and A_2 as well as $B(t,0) \neq 0$ for solutions of the system (1.7) with $x_0 \in \mathbb{R}^n$, $\|x_0\| < \infty$, we have (1.9).

When the inequality (1.17) is satisfied, the assertion in Theorem 1.1 follows directly from the estimate (1.17). □

Corollary 1.1. (*see* [15]). *If:* (1) *in the system* (1.7) $\lim_{t\to\infty} A(t) = A$, *and all solutions of the system*

$$\frac{dx}{dt} = Ax \tag{1.18}$$

are bounded; (2) *the matrix* $B(t, x) = \bar{B}(t)$ *is continuous for all* $t \geq t_0$ *and* $x \in \mathbb{R}^n \backslash \{0\}$ *and such that* $\int_{t_0}^{\infty} \|\bar{B}(s)\| ds < \infty$, *then all solutions of the system*

$$\frac{dy}{dt} = (A + \bar{B}(t))y \tag{1.19}$$

are bounded.

Definition 1.2. The solution $x = 0$ of the system (1.7) is (ε, δ, J)-stable if, for any $\varepsilon > 0$ and $t_0 \in \mathbb{R}_+$, there exists $\delta(t_0, \varepsilon) > 0$ such that if $\|x(t_0)\| < \delta$, then

$$\|x(t)\| < \varepsilon,$$

where $x(t) = x(t, t_0, x_0)$ is the solution of the initial problem (1.7)–(1.8) for all $t \in J$.

When studying the stability of the zero solution of the system (1.7), we assume that $B(t, x) = 0$ for $x = 0$.

The following assertion holds.

Theorem 1.2. *Let the system* (1.7) *satisfy the conditions of assumptions* A_1 *and* A_2, *the inequality* (1.10) *holds for all* $t \in J$, *and, in addition, for any* $\varepsilon > 0$ *and* $t_0 \in \mathbb{R}_+$, *there exist* $\delta > 0$ *such that*

$$\exp\left(\int_{t_0}^{t} a_1(s)ds\right)(1 - \Phi(t_0, t))^{-1} < \frac{\varepsilon}{\delta} \tag{1.20}$$

for all $t \in J$.

Then, the zero solution of the system (1.7) *is* (ε, δ, J)-*stable.*

The proof is similar to that of Theorem 1.1.

Corollary 1.2. *In the system* (1.7), *let* $A(t) = 0$ *for all* $t \geq t_0$ *and the assumption of condition* A_2 *be satisfied. If for any* $\varepsilon > 0$ *and* $t_0 \in \mathbb{R}_+$, *there exists* $\delta(t_0, \varepsilon) > 0$ *such that*

$$\left(1 - \delta \int_{t_0}^{t} a_2(s)ds\right)^{-1} < \frac{\varepsilon}{\delta}$$

holds for all $t \in J$, as soon as for $\|x_0\| < \delta$, the inequality

$$\delta \int_{t_0}^t a_2(s)ds < 1$$

for all $t \in J$, then the zero solution of the system

$$\frac{dx}{dt} = B(t, x)x \tag{1.21}$$

is stable in the sense of Definition 1.2.

Proof. We write the system of equations (1.21) in the integral form

$$x(t) = x_0 + \int_{t_0}^t B(s, x(s))x(s)ds. \tag{1.22}$$

□

If condition A_2 is satisfied, from (1.22), we get

$$\|x(t)\| \le \|x_0\| + \int_{t_0}^t a_2(s)\|x(s)\|^2 ds. \tag{1.23}$$

From the inequality (1.23), it follows that if

$$\|x_0\| \int_{t_0}^t a_2(s)ds < 1 \quad \text{for all } t \in J,$$

then we have the estimate

$$\|x(t)\| \le \|x_0\| \left(1 - \|x_0\| \int_{t_0}^t a_2(s)ds\right)^{-1} \tag{1.24}$$

for all $t \in J$. Under the conditions of Corollary 1.2, the estimate (1.24) implies that the zero solution of the system (1.21) is (ε, δ, J)-stable.

1.2.4   Bilinear system with autonomous linear approximation

Consider a bilinear system in the form

$$\frac{dx}{dt} = (A + B(t, x))x, \quad x(t_0) = x_0, \tag{1.25}$$

where $x \in \mathbb{R}^n$, $B(t, x)$ is continuous for all $t \ge 0$ and $\|x\| \le h$, such that

$$\|B(t, x)\| \le a_2(t)\|x\|. \tag{1.26}$$

where $a_2(t)$ is a function from assumption A_2.

Suppose that all solutions of the linear approximation of the system (1.25)

$$\frac{dx}{dt} = Ax \qquad (1.27)$$

are bounded. We write the solution $x(t)$ of the system of equations (1.25) as

$$x(t) = \exp(At)x_0 + \int_0^t \exp(A(t-s))\, B(s, x(s))x(s)ds. \qquad (1.28)$$

The boundedness of the solutions of the system (1.27) implies the existence of the constant $K > 0$ such that

$$\| \exp(At) \| \leq K \qquad (1.29)$$

for all $t \geq 0$. Given assumption A_2 and the estimate (1.29), from the relation (1.28), we obtain the inequality

$$\|x(t)\| \leq K\|x_0\| + \int_0^t K a_2(s)\|x(s)\|^2 ds. \qquad (1.30)$$

Applying Lemma 1.4 to the inequality (1.30), we get the estimate

$$\|x(t)\| \leq K\|x_0\| \exp\left(\int_0^t K a_2(s)\|x(s)\| ds \right). \qquad (1.31)$$

Performing simple transformations of the inequality (1.31), we arrive at the estimate

$$\|x(t)\| \leq \frac{K\|x_0\|}{1 - K^2\|x_0\| \int_0^t a_2(s)ds} \qquad (1.32)$$

for all $t \in J$, for which

$$K^2\|x_0\| \int_0^t a_2(s)ds < 1. \qquad (1.33)$$

Based on the estimate (1.32), we formulate conditions for β-*boundedness* and (ε, δ, J)-*stability* of the zero solution of the system (1.25).

Theorem 1.3. *Let the system* (1.25) *satisfy the following conditions:*

(1) *all solutions of the system* (1.27) *are bounded;*
(2) *the inequality* (1.26) *is fulfilled;*
(3) *for the initial values* $0 < \|x_0\| < \infty$, *the inequality* (1.33) *is satisfied for all* $t \in J$;

(4) *for a given constant value $\beta > 0$, the inequality*

$$\left(1 - K^2\|x_0\| \int_0^t a_2(s)ds\right)^{-1} < \frac{\beta}{K\|x_0\|} \qquad (1.34)$$

is fulfilled for all $t \in J$.

Then, all solutions of the system (1.34) *are β-bounded.*

The proof of this theorem follows directly from the estimate (1.32) and the inequality (1.34).

Theorem 1.4. *Let the system* (1.25) *satisfy conditions* (1) *and* (2) *of Theorem 1.3 and, moreover,*
(3) *for any $\varepsilon > 0$, there exists $\delta = \delta(\varepsilon) > 0$ such that when $\|x_0\| < \delta(\varepsilon)$,*

$$K^2\delta(\varepsilon) \int_0^t a_2(s)ds < 1 \quad \text{for all } t \in J;$$

(4)

$$\left(1 - K^2\delta(\varepsilon) \int_0^t a_2(s)ds\right)^{-1} < \frac{\varepsilon}{K\delta}$$

for all $t \in J$.
Then, the zero solution of the system (1.25) *is (ε, δ, J)-stable.*

The proof of Theorem 1.4 is similar to that of Theorem 1.2.

1.3 Estimates for Solutions of Quasi-Linear Systems

In this section, the method for *estimating* the *norms of solutions* proposed in Section 1.2 is extended to quasi-linear non-autonomous systems. Based on the estimates obtained, sufficient conditions are established for the *β-boundedness* and *(ε, δ, J-stability)* of the motion of such a class of systems of perturbed motion equations.

1.3.1 Statement of the problem

We consider a *quasi-linear system* of differential equations of perturbed motion

$$\frac{dx}{dt} = A(t)x + f(t, x), \quad x(t_0) = x_0. \qquad (1.35)$$

Here, $x \in \mathbb{R}^n$, $f \in C(\mathbb{R}_+ \times \mathbb{R}^n, \mathbb{R}^n)$, and $A(t)$ is an $n \times n$ matrix with elements continuous on any finite interval. We assume that the solution $x(t) = x(t, t_0, x_0)$ of the initial problem (1.35) exists and is unique for all $0 \le t < \infty$ and $(t_0, x_0) \in \mathbb{R}_+ \times \mathbb{R}^n$.

Equations of the form (1.35) occur in the description of many processes and phenomena in mechanics (see, for example, [34, 53] and the bibliography therein). In addition, these equations can be used to describe the motion of a linear system,

$$\frac{dx}{dt} = A(t)x, \quad x(t_0) = x_0, \qquad (1.36)$$

under perturbations if, for the vector function $f \in C(\mathbb{R}_+ \times \mathbb{R}^n, \mathbb{R}^n)$, only norm estimates are known.

In order to obtain sufficient conditions for the boundedness and/or stability of the solutions of the system (1.35), it is necessary to estimate the norm of the solutions of the equations under consideration under various assumptions about the properties of the solutions of the system (1.36) and the vector function of nonlinearities in (1.35).

Further, the necessary estimates for the norms of solutions will be obtained using the method of integral nonlinear inequalities (see [24, 45]).

1.3.2   Generalised Brauer estimate

First, we obtain a *Brauer estimate* for the norm of solutions $x(t)$ of the initial problem (1.35) under the following assumptions:

A_3. For all $t \ge t_0$, there exists a non-negative integrable function $b(t)$ such that

$$\|A(t)\| \le b(t) \quad \text{for all} \quad t \ge t_0.$$

A_4. For all $t \ge t_0$ and $u \ge 0$, there exists a continuous integrable function $w(t, u)$, $w(t, 0) = 0$, such that (cf. [24])

$$\|f(t, x)\| \le w(t, \|x\|)$$

for all $(t, x) \in \mathbb{R}_+ \times \mathbb{R}^n$.

Lemma 1.5. *Let the system* (1.35) *satisfy the conditions of assumptions A_3 and A_4. Then, for any solution $x(t) = x(t, t_0, x_0)$ with the initial conditions $x_0 \colon \|x_0\| \le c$, $0 < c < +\infty$, the following inequality holds:*

$$\|x(t)\| \le c + \int_{t_0}^{t} [b(s)\|x(s)\| + w(s, \|x(s)\|)]\, ds \qquad (1.37)$$

for all $t \ge t_0 \ge 0$.

If there exists:

(a) *a continuous and non-negative function $v(t)$ for all $t \geq t_0$;*
(b) *a continuous non-negative and non-decreasing function $g(u)$ for $u \geq 0$ such that*

$$w\left(t,\, z\exp\left(\int_{t_0}^{t} b(s)\, ds\right)\right)\exp\left(-\int_{t_0}^{t} b(s)\, ds\right) \leq v(t)g(z),$$

$$t \geq t_0, \quad z \geq 0,$$

then for all $t \in [t_0, \beta)$, the following inequality holds:

$$\|x(t)\| \leq G^{-1}\left[G(c) + \int_{t_0}^{t} v(s)\, ds\right]\exp\left(\int_{t_0}^{t} b(s)\, ds\right), \tag{1.38}$$

where G^{-1} is the inverse function with respect to the function $G(u)$:

$$G(u) - G(u_0) = \int_{u_0}^{u} \frac{ds}{g(s)}, \quad 0 < u_0 \leq c \leq u \leq \infty,$$

and the value of β is determined from the relation

$$\beta = \sup\left\{t \geq t_0\colon G(c) + \int_{t_0}^{t} v(s)\, ds \in \operatorname{dom} G^{-1}\right\}.$$

(c) *If, in addition to conditions (a) and (b), there exists a constant $a^0 > 0$ such that*

$$\int_{t_0}^{\infty} v(t)\, dt \leq \int_{a^0}^{\infty} \frac{ds}{g(s)},$$

then the inequality (1.38) holds for all $t \geq t_0$, i.e., $\beta = \infty$ for the values $c \in (0,\, a^0)$.

Proof. We equate the right-hand side of (1.37) to the expression $p(t)\exp\left(\int_{t_0}^{t} b(s)ds\right)$ with some function $p(t)$. Using the inequality (1.38) and condition (b) of Lemma 1.5, it is easy to obtain the estimate

$$\left[\frac{dp}{dt} + b(t)p(t)\right]\exp\left(\int_{t_0}^{t} b(s)\, ds\right) = b(t)\|x(t)\| + w(t, \|x(t)\|)$$

$$\leq \left[b(t)p(t) + v(t)g\left(\|x(t)\|\exp\left(-\int_{t_0}^{t} b(s)\, ds\right)\right)\right]$$

$$\times \exp\left(\int_{t_0}^{t} b(s)\, ds\right).$$

Since the function g is non-decreasing and

$$\|x(t)\| \leq p(t) \exp\left(\int_{t_0}^{t} b(s)\, ds\right),$$

we arrive at the inequality

$$\frac{dp}{dt} \leq v(t)g(p(t)), \quad p(t_0) = c.$$

Hence, by the *Bihari lemma* (see [19]), we obtain

$$p(t) \leq G^{-1}\left[G(c) + \int_{t_0}^{t} v(s)\, ds\right]$$

for all $t \in (t_0, \beta)$. This proves (1.38). □

To prove the second assertion of Lemma 1.5, note that the continuity condition for the function $p(t)$ is the inequality

$$G(c) + \int_{t_0}^{\infty} v(s)\, ds \leq \int_{u_0}^{\infty} \frac{ds}{g(s)},$$

or

$$\int_{t_0}^{\infty} v(s)\, ds \leq -\int_{u_0}^{c} \frac{ds}{g(s)} + \int_{u_0}^{\infty} \frac{ds}{g(s)} = \int_{c}^{\infty} \frac{ds}{g(s)}.$$

This inequality holds for any value of $c \in (0, a^0)$ for which condition (c) of Lemma 1.5 is satisfied. Since $c < a^0$, we get

$$\int_{t_0}^{\infty} v(s)\, ds \leq \int_{a_0}^{\infty} \frac{ds}{g(s)} < \int_{c}^{\infty} \frac{ds}{g(s)}.$$

Therefore, for all $c \in (0, a^0)$, the value of the constant is $\beta = \infty$. This proves Lemma 1.5.

1.3.3   A special case of nonlinearity estimation

Let us continue with the consideration of the nonlinear system of differential equations of perturbed motion in the form (1.35). The properties of stability and boundedness of solutions to the system (1.35) are often studied by comparing them with the properties of solutions to the equations (1.36) (see, for example, [35, 141]).

Let us first establish an *estimate* for the *norm of solutions* $x(t)$ of the system (1.35) with assumption A_4 replaced by the following:

A_5. There exist a non-negative integrable function $c(t)$ for all $t \geq t_0 \geq 0$ and a constant $\alpha > 1$ such that

$$\|f(t, x)\| \leq c(t)\|x\|^{\alpha}$$

for all $(t, x) \in \mathbb{R}_+ \times D$, where $D = \{x \in \mathbb{R}^n : \|x\| \leq d\}$.

Lemma 1.6. *Let the system of equations (1.35) satisfy the conditions of assumptions A_3 and A_5. Then, the norm of the solution $x(t) = x(t, t_0, x_0)$ satisfies the estimate*

$$\|x(t)\| \leq \|x_0\| \exp \int_{t_0}^{t} b(s)ds (1 - L(t_0, t))^{-\frac{1}{\alpha - 1}} \qquad (1.39)$$

for all $t \in J = [t_0, T)$, where $t_0 \in \mathbb{R}_+$, as soon as

$$L(t_0, t) = (\alpha - 1)\|x_0\|^{\alpha - 1} \int_{t_0}^{t} c(s) \exp\left((\alpha - 1)\int_{t_0}^{s} b(\tau)d\tau\right) ds < 1. \quad (1.40)$$

Proof. Let $x(t)$ be the solution of the system of equations (1.35) with the initial values $x(t_0) = x_0$ and $t_0 \geq 0$. From the equation (1.40), under conditions A_4 and A_5, we obtain an estimate for the norm of the solution $x(t)$ in the form

$$\|x(t)\| \leq \|x_0\| + \int_{t_0}^{t} b(s)\|x(s)\|ds + \int_{t_0}^{t} c(s)\|x(s)\|^{\alpha} ds. \qquad (1.41)$$

$\square$

The inequality (1.41) is transformed into a pseudo-linear form,

$$\|x(t)\| \leq \|x_0\| + \int_{t_0}^{t} \left(b(s) + c(s)\|x(s)\|^{\alpha - 1}\right) \|x(s)\|ds, \qquad (1.42)$$

and applying Lemma 1.2, we obtain the estimate

$$\|x(t)\| \leq \|x_0\| \exp\left(\int_{t_0}^{t} \left(b(s) + c(s)\|x(s)\|^{\alpha - 1}\right) ds\right) \qquad (1.43)$$

for all $t \geq t_0 \geq 0$. Denote $\|x(t)\| = \psi(t)$ for all $t \geq t_0$, and rewrite the inequality (1.43) as

$$\psi^{\alpha - 1}(t) \leq \|x_0\|^{\alpha - 1} \exp\left[(\alpha - 1)\int_{t_0}^{t} \left(b(s) + c(s)\psi^{\alpha - 1}(s)\right) ds\right]. \qquad (1.44)$$

Multiplying both sides of (1.44) by the expression

$$-(\alpha - 1)c(s) \exp\left(-(\alpha - 1)\int_{t_0}^{t} c(s)\psi^{\alpha - 1}(s)ds\right),$$

we get the estimate

$$-(\alpha - 1)c(t)\psi^{\alpha - 1}(t) \exp\left[-(\alpha - 1)\int_{t_0}^{t} c(s)\psi^{k - 1}(s)ds\right]$$

$$\geq -(\alpha - 1)\|x_0\|^{\alpha - 1}c(t) \exp\left[(\alpha - 1)\int_{t_0}^{t} b(s)ds\right],$$

from which it follows that

$$-(\alpha - 1)\|x_0\|^{\alpha-1}c(t)\exp\left[(\alpha - 1)\int_{t_0}^s b(s)ds\right]$$

$$\leq \frac{d}{dt}\left[\exp\left(-(\alpha - 1)\int_{t_0}^t c(s)\psi^{\alpha-1}(s)ds\right)\right].$$

Taking $\exp^{t_0} = 1$ and integrating this inequality between t_0 and t, we find

$$1 - (\alpha - 1)\|x_0\|^{\alpha-1}\int_{t_0}^t c(s)\exp\left[(\alpha - 1)\int_{t_0}^s b(\tau)d\tau\right]ds$$

$$\leq \exp\left[-(\alpha - 1)\int_{t_0}^s c(s)\psi^{\alpha-1}(s)ds\right]. \tag{1.45}$$

When the condition (1.40) is satisfied, we get

$$\exp\left[(\alpha - 1)\int_{t_0}^s c(s)\psi^{\alpha-1}(s)ds\right]$$

$$\leq (1 - L(t_0, t))^{-1}.$$

In this case, the inequality (1.44) takes the form

$$\psi^{\alpha-1}(t) \leq \|x_0\|^{\alpha-1}\exp\left[(\alpha - 1)\int_{t_0}^t b(s)ds\right](1 - L(t_0, t))^{-1},$$

which implies an estimate for the norm of the solution $\|x(t)\|$ in the form (1.39) for all $t \in J = [t_0, T)$.

Lemma 1.6 is proved.

Remark 1.1. Here, to estimate the interval $J = [t_0, T)$, we use the formula

$$T \leq \sup\{t \in \mathbb{R}_+ : L(t_0, t) < 1\}.$$

The estimate (1.39) allows us to establish conditions for *β-boundedness* and (ε, δ, J)-*stability* of the solution to the system (1.35) in the sense of the definitions given as follows.

Note that the monograph [169] contains classical definitions of various types of boundedness and stability of motion.

When studying the boundedness of solutions of the system (1.35), we assume that $f(t, 0) \neq 0$ for all $t \in \mathbb{R}_+$.

Let us define the properties of the solutions, taking into account the boundedness of the interval on which their behaviour is considered.

Definition 1.3. The solution $x(t)$ of the system (1.35) is (ρ, β)-*bounded* if, for a given $\beta > 0$, for any $\rho > 0$ and $t_0 \in \mathbb{R}_+$ such that as soon as $\|x(t_0)\| < \rho$, the inequality

$$\|x(t)\| < \beta$$

holds for all $t \in J$.

The following assertions hold.

Theorem 1.5. *Let assumptions A_4 and A_5 and the conditions of Lemma 1.6 hold for all $t \in J$, $0 < \|x_0\| \le \infty$. If a given constant $\beta > 0$ satisfies the condition*

$$\exp \int_{t_0}^{t} b(s)ds \left(1 - L(t_0, t)\right)^{-\frac{1}{\alpha - 1}} \le \frac{\beta}{\|x_0\|}$$

for all $t \in J$, then the solution $x(t)$ of the system (1.35) is β-bounded.

Theorem 1.6. *Let assumptions A_4 and A_5 and the conditions of Lemma 1.6 hold for all $t \in J$, $0 < \|x_0\| \le \rho$. If for a given constant $\beta > 0$, the inequality*

$$\exp \int_{t_0}^{t} b(s)ds \left(1 - L(t_0, t)\right)^{-\frac{1}{\alpha - 1}} \le \frac{\beta}{\rho}$$

holds for all $t \in J$, then the solution $x(t)$ of the system (1.35) is (ρ, β)-bounded.

The proofs of Theorems 1.5 and 1.6 follow directly from the estimate for the norm of solutions $x(t)$ of the system (1.35) in the form (1.39) under appropriate initial conditions.

The following definition takes into account the finiteness of the time interval on which the stability of the system's motion is considered.

Further, when considering the problem of the stability of the state $x = 0$ of the system (1.35), it is assumed that $f(t, x) = 0$ for $x = 0$ and for all $t \in \mathbb{R}_+$.

The following assertion holds.

Theorem 1.7. *Let the conditions of assumptions A_4 and A_5 and Lemma 1.6 hold for all $(t, x) \in \mathbb{R}_+ \times D$. If for any $\varepsilon > 0$ and $t_0 \ge 0$, there exists $\delta(t_0, \varepsilon) > 0$ such that for $\|x_0\| < \delta(t_0, \varepsilon)$, the estimate*

$$\exp \int_{t_0}^{t} b(s)ds \left(1 - L(t_0, t)\right)^{-\frac{1}{\alpha - 1}} \le \frac{\varepsilon}{\delta}$$

holds for all $t \in J$, then the solution $x(t) = 0$ of the system (1.35) is (ε, δ, J)-stable.

The proof of Theorem 1.7 follows directly from the estimate for the norm of solutions $x(t)$ in the form (1.39).

Corollary 1.3. *Let the linear approximation matrix $A(t) \equiv 0$ for all $t \geq t_0 \geq 0$ be in the system of perturbed motion equations (1.35). We consider the system of equations*

$$\frac{dx}{dt} = f(t, x), \quad x(t_0) = x_0. \tag{1.46}$$

This is an essentially nonlinear system *since there is no linear approximation in it.*

Such equations are encountered when considering mechanical systems with dry friction in the theory of electroacoustic waveguides and in other problems. Systems with sector nonlinearity are close to this type of systems.

Assume that condition A_5, for the right-hand side of the equation (1.46), is satisfied with a continuous and non-negative function $c(t)$.

Then, we have the estimate

$$\|x(t)\| \leq \|x_0\| + \int_{t_0}^{t} c(s)\|x(s)\|^\alpha ds$$

for all $t \geq t_0 \geq 0$.

Applying to this inequality the same technique of analysis that was used in the proof of Lemma 1.6, we obtain the estimate

$$\|x(t)\| \leq \frac{\|x_0\|}{\left(1 - (\alpha - 1)\|x_0\|^{\alpha-1} \int_{t_0}^{t} c(s)ds\right)^{\frac{1}{\alpha-1}}} \tag{1.47}$$

for all $t \in J$ as soon as the following inequality is satisfied:

$$(\alpha - 1)\|x_0\|^{\alpha-1} \int_{t_0}^{t} c(s)\, ds < 1$$

for all $t \in J$.

Remark 1.2. Here, to estimate the interval $J = [t_0, T)$, we use the formula

$$T \leq \sup\left\{ t \in \mathbb{R}_+ : (\alpha - 1)\|x_0\|^{\alpha-1} \int_{t_0}^{t} c(s)\, ds < 1 \right\}.$$

Remark 1.3. The estimate (1.47) is also obtained by applying Bihari's inequality (see [19]) to the inequality

$$\|x(t)\| \leq \|x_0\| + \int_{t_0}^{t} c(s)\|x(s)\|^{\alpha} ds,$$

with the function $G(u) = \|x\|^{\alpha}$, $\alpha > 0$, $\alpha \neq 1$.

Based on the estimate (1.47) of the norm of solutions of the system (1.46), one can obtain sufficient conditions for (ρ, β)-boundedness and (ε, δ, J)-stability in the same way as in Theorems 1.5–1.7.

1.4 Boundedness with Respect to Some Variables

The purpose of this section is to obtain sufficient conditions for boundedness of motion with respect to some variables on the basis of an estimate of the *norm of solutions* of *coupled equations* of perturbed motion. One of the applications of the obtained result is the analysis of solutions of systems of differential equations with fast and slow variables (see [23, 39] and the bibliography therein).

1.4.1 Formulation of the problem

We consider the system of differential equations of perturbed motion

$$\frac{dx}{dt} = f(t, x, y),$$

$$\frac{dy}{dt} = g(t, x, y), \qquad (1.48)$$

$$x(t_0) = x_0, \quad y(t_0) = y_0,$$

where $x \in \mathbb{R}^n$, $y \in \mathbb{R}^m$, and f and g are the continuous vector functions on the set $\mathbb{R}_+ \times \mathbb{R}^n \times \mathbb{R}^m$. It is assumed that the solutions of the initial problem (1.48) are y-continuable, i.e., any solution $(x^T(t), y^T(t))^T$ of the system (1.48) is defined for all $t \geq t_0$ for which $\| y(t) \| \leq H$, where $H > 0$ is some constant value.

Next, we introduce the notation

$$\|x\| = \left(\sum_{i=1}^{n} x_i^2 \right)^{\frac{1}{2}}, \quad \|y\| = \left(\sum_{j=1}^{m} y_i^2 \right)^{\frac{1}{2}},$$

$$\|z\| = \left(\sum_{i=1}^{n} x_i^2 \right)^{\frac{1}{2}} + \left(\sum_{j=1}^{m} y_i^2 \right)^{\frac{1}{2}} = (\|x\|^2 + \|y\|^2)^{\frac{1}{2}},$$

and we provide definitions necessary for further presentation of the results.

Note that the classical definition of motion stability with respect to some variables goes back to A. M. Lyapunov (see [76]). The development of this idea of Lyapunov was carried out by many scientists, and the results obtained are summarised in a number of monographs, including [148, 176]. The definition given in the following takes into account the finiteness of the interval on which the motion of the system is considered.

Definition 1.4. The motion $(x(t), y(t))^T$ of the system (1.48) is:

(a) *x-bounded* if for any $t_0 \in \mathbb{R}_+$ and $z_0 \in \mathbb{R}^n \times \mathbb{R}^m$, $\|z_0\| < \infty$, given a constant $\beta(t_0) > 0$, $\|x(t, t_0, z_0)\| < \beta(t_0)$ for all $t \in J$, where $x(t, t_0, z_0)$ is the solution of the initial problem (1.48);

(b) *x-bounded uniformly in t_0* if in Definition 1.4 for any $z_0 \in \mathbb{R}^n \times \mathbb{R}^m$, $\|z_0\| < \infty$, given $\beta > 0$ independent of t_0, $\|x(t, t_0, z_0)\| < \beta$ for all $t \in J$;

(c) *x-bounded uniformly in y_0* if for any $t_0 \geq 0$ and a compact set $K \subset \mathbb{R}^n \times \mathbb{R}^m$, there exists $\beta^*(t_0, K) > 0$ such that $z_0 \in K$ implies $\|x(t, t_0, z_0)\| < \beta^*(t_0, K)$ for all $t \in J$;

(d) *x-bounded uniformly in $\{t_0, z_0\}$* if $\beta > 0$ can be chosen in Definition 1.4(a) for any compact set $K^* > 0$ independent of t_0.

It is of interest to consider the problem of sufficient conditions for the x-boundedness of the motion of the system (1.48) in the sense of Definition 1.4.

1.4.2   Estimations of solutions of the system (1.48)

From the equations (1.48), it follows that

$$x(t) = x_0 + \int_{t_0}^{t} f(s, x(s), y(s))ds,$$

$$y(t) = y_0 + \int_{t_0}^{t} g(s, x(s), y(s))ds, \quad t \geq t_0. \tag{1.49}$$

From here, we find that

$$\|x(t)\| + \|y(t)\| \leq \|x_0\| + \|y_0\| + \int_{t_0}^{t} (\| f(s, x(s), \ y(s)) \|$$

$$+ \| g(s, x(s), y(s)) \|)ds, \quad t \geq t_0. \tag{1.50}$$

Assume that the following conditions are fulfilled for the system (1.48):

A_6. There exist non-negative integrable functions $\psi_1(t)$ and $\psi_2(t)$ such that in the range $(t, x, y) \in \mathbb{R}_+ \times \mathbb{R}^n \times \mathbb{R}^m$, the following estimates are satisfied:

(a) $\| f(t, x, y) \| \le \psi_1(t)(\|x\| + \|y\|)^p,$

(b) $\| g(t, x, y) \| \le \psi_2(t)(\|x\| + \|y\|)^q,$ (1.51)

where $1 < p < q < \infty$.

Denote $\|z(t)\| = \|x(t)\| + \|y(t)\|$, and show that the following assertion holds.

Lemma 1.7. *Assume that the system of equations* (1.48) *satisfies the conditions of assumption* A_6. *Then,*

$$\|z(t)\| \le \|z_0\| \left(1 - N(t_0, t)\right)^{-\frac{1}{p+q-2}}$$ (1.52)

for all $t \in J$, *for which*

$$N(t_0, t) = (p + q - 2)\Big[\|z_0\|^{p-1} \int_{t_0}^{t} \psi_1(s) ds + \|z_0\|^{q-1}$$

$$\times \int_{t_0}^{t} \psi_2(s) ds\Big] < 1.$$ (1.53)

Proof.   Taking into account the notation $\|z(t)\|$, from the inequality (1.51), we obtain the estimate

$$\|z(t)\| \le \|z_0\| + \int_{t_0}^{t} \left(\psi_1(s)\|z(s)\|^p + \psi_2(s)\|z(s)\|^q\right) ds$$ (1.54)

and, further,

$$\|z(t)\| \le \|z_0\| + \int_{t_0}^{t} \left(\psi_1(s)\|z(s)\|^{p-1} + \psi_2(s)\|z(s)\|^{q-1}\right) \|z(s)\| ds, \quad t \ge t_0.$$ (1.55)

$\square$

Applying Lemma 1.2 to the inequality (1.55), we obtain the estimate (cf. [124])

$$\|z(t)\| \le \|z_0\| \exp\left[\int_{t_0}^{t} (\psi_1(s)\|z(s)\|^{p-1} + \psi_2(s)\|z(s)\|^{q-1}) ds\right]$$ (1.56)

for all $t \ge t_0$. It follows from the estimate (1.56) that

$$\|z(t)\|^{p-1} \le \|z_0\|^{p-1} \exp\left[(p-1) \int_{t_0}^{t} (\psi_1(s)\|z(s)\|^{p-1}\right.$$

$$\left. + \psi_2(s)\|z(s)\|^{q-1}) ds\right]$$

$$\text{and} \quad \|z(t)\|^{q-1} \le \|z_0\|^{q-1} \exp\left[(q-1) \int_{t_0}^{t} (\psi_1(s)\|z(s)\|^{p-1}\right.$$ (1.57)

$$\left. + \psi_2(s)\|z(s)\|^{q-1}) ds\right]$$

for all $t \geq t_0$. Let $p > 1$ and $q > 1$. Then, (1.57) implies the estimates

$$\|z(t)\|^{p-1} \leq \|z_0\|^{p-1} \exp\left[(p+q-2)\int_{t_0}^{t}(\psi_1(s)\|z(s)\|^{p-1}\right.$$
$$\left. + \psi_2(s)\|z(s)\|^{q-1})ds\right],$$

$$\|z(t)\|^{q-1} \leq \|z_0\|^{q-1} \exp\left[(p+q-2)\int_{t_0}^{t}(\psi_1(s)\|z(s)\|^{p-1}\right.$$
$$\left. + \psi_2(s)\|z(s)\|^{q-1})ds\right]$$

(1.58)

for all $t \geq t_0$.

Multiplying both sides of the first inequality from (1.58) by $-(p+q-2)\psi_1(t)$ and the second by $-(p+q-2)\psi_2(t)$, we get

$$-\|z(t)\|^{p-1}\psi_1(t)(p+q-2)\exp\left[-(p+q-2)\int_{t_0}^{t}(\psi_1(s)\|z(s)\|^{p-1}\right.$$
$$\left. + \psi_2(s)\|z(s)\|^{q-1})ds\right] \geq \|z_0\|^{p-1}(p+q-2)\psi_1(t),$$

$$-\|z(t)\|^{q-1}\psi_2(t)(p+q-2)\exp\left[-(p+q-2)\int_{t_0}^{t}(\psi_1(s)\|z(s)\|^{p-1}\right.$$
$$\left. + \psi_2(s)\|z(s)\|^{q-1})ds\right] \geq -\|z_0\|^{q-1}(p+q-2)\psi_2(t)$$

(1.59)

for all $t \geq t_0$. It is easy to see that

$$\frac{d}{dt}\exp\left[-(p+q-2)\int_{t_0}^{t}(\psi_1(s)\|z(s)\|^{p-1} + \psi_2(s)\|z(s)\|^{q-1})ds\right]$$
$$\geq -\|z_0\|^{p-1}(p+q-2)\psi_1(t) - \|z_0\|^{q-1}(p+q-2)\psi_2(s)$$

(1.60)

for all $t \geq t_0$. Integrating both sides of the inequality (1.60) between t_0 and t, we obtain the estimate

$$\exp\left[(p+q-2)\int_{t_0}^{t}(\psi_1(s)\|z(s)\|^{p-1}\right.$$
$$\left. + \psi_2(s)\|z(s)\|^{q-1})ds\right] \leq (1 - N(t_0, t))^{-1}$$

(1.61)

for those values of $t \in J$, for which $N(t_0, t) < 1$.

In view of the estimate (1.61), it follows from the inequality (1.58) that

$$\|z(t)\|^{p+q-2}\|z_0\|^{-(p+q-2)} \leq (1 - N(t_0, t))^{-1},$$

or, finally,

$$\|z(t)\| \le \|z_0\|(1 - N(t_0, t))^{-\frac{1}{p+q-2}} \tag{1.62}$$

for all $t \in J$, for which $N(t_0, t) < 1$. Lemma 1.7 is proved.

Lemma 1.7 has a corollary that takes into account the structure of the right-hand side of the system (1.48).

Corollary 1.4. *Let, under assumption A_6 (inequality (b)), the magnitude $q = 1$. Then, if*

$$N_1(t_0, t) = (p - 1)\|z_0\|^{p-1} \int_{t_0}^{t} \psi_1(s) \exp$$

$$\times \left[(p - 1) \int_{t_0}^{s} \psi_2(\tau)d\tau \right] ds < 1 \tag{1.63}$$

for all $t \in J \subset \mathbb{R}_+$. Then,

$$\|z(t)\| \le \|z_0\| \exp \left[\int_{t_0}^{t} \psi_2(s)ds \right] (1 - N_1(t_0, t))^{-\frac{1}{p-1}}$$

for all $t \in J$.

The *proof* of this assertion is similar to that of Lemma 1.6.

1.4.3 x-boundedness of motion conditions

Lemma 1.7 and its Corollary 1.4 allow us to establish new conditions for x-boundedness of solutions of the system (1.48).

Theorem 1.8. *Assume that the system (1.48) satisfies all conditions of Lemma 1.7 and, in addition:*

(1) $N(t_0, t) < 1$ *for all $t \in J$;*
(2) *there exists a constant $\lambda > 0$, and for a given value $\beta > 0$, the inequality*

$$\frac{1}{\lambda} (1 - N(t, t_0))^{-\frac{1}{p+q-2}} < \frac{\beta}{\|z_0\|}$$

holds for all $t \in J$ and $\|z_0\| < \infty$.
Then, the motion of the system (1.48) is x-bounded.

Proof. Let $z(t) = (x^T(t), y^T(t))^T$ be the solution of the system (1.48) with the initial conditions $t_0 \in J$ and $z_0 \in \mathbb{R}^n \times \mathbb{R}^m$, $\|z_0\| < \infty$. □

Suppose that under the conditions of Theorem 1.8, for some $t_1 \in J$, $t_1 > t_0$, the solution x of the system (1.48) satisfies the relation $\|x(t_1, t_0, z_0)\| = \beta$, where $\beta > 0$ is some given value.

Under the conditions of Lemma 1.7 and condition (1) of Theorem 1.7, we have the estimate

$$\|z(t)\| < \|z_0\| \left(1 - N(t_0, t)\right)^{-\frac{1}{p+q-2}}$$

for all $t \in [t_0, t_1)$. Since $\lambda \|x\| \leq \|x\| + \|y\|$ for some $\lambda > 0$,

$$\lambda \|x(t_1)\| \leq \|z(t_1)\| \leq \|z_0\| \left(1 - N(t_0, t)\right)^{-\frac{1}{p+q-2}},$$

and by virtue of condition (2), we have

$$\|x(t_1)\| < \frac{\|z_0\|\beta}{\|z_0\|} = \beta.$$

This contradicts the assumption that $t_1 > t_0$ exists for which $\|x(t_1)\| = \beta$. Hence, $\|x(t)\| < \beta$ for all $t \in J$. Theorem 1.8 is proved.

Theorem 1.9. *Assume that the system (1.48) satisfies all the conditions of Lemma 1.7 and, in addition:*

(1) *$N(t_0, t) < 1$, uniformly in $t_0 \in J$ for all $t \in J$;*
(2) *there exists a constant $\lambda > 0$, and for a given value $\beta > 0$, the inequality*

$$\frac{1}{\lambda} \left(1 - N(t_0, t)\right)^{-\frac{1}{p+q-2}} < \frac{\beta}{\|z_0\|}$$

holds for all $t \geq t_0$.

Then, the motion of the system (1.48) is x-bounded uniformly in t_0.

The proof of Theorem 1.8 is similar to that of Theorem 1.8.

Next, we consider the compact set $K \subset \mathbb{R}^n \times \mathbb{R}^m$, for example, $K = \{(x_0, y_0) \in \mathbb{R}^n \times \mathbb{R}^m : \|z_0\|^2 \leq \alpha^2\}$, where $\alpha > 0$ is some constant value.

Theorem 1.10. *Assume that the system (1.48) satisfies all the conditions of Lemma 1.7 and, in addition:*

(1) *for any compact set K, the condition*

$$N^*(t_0, t) = (p + q - 2) \left[\alpha^{2(p-1)} \int_{t_0}^{t} \psi_1(s)ds + \alpha^{2(q-1)} \int_{t_0}^{t} \psi_2(s)ds\right] < 1$$

is satisfied for all $t \in J$;

(2) *there exists* $\lambda > 0$ *and, given* $\beta^* > 0$, *the inequality*

$$\frac{1}{\lambda}\left(1 - N^*(t_0, t)\right)^{-\frac{1}{p+q-2}} < \frac{\beta^*}{\alpha^2}$$

holds for all $t \in J$.

Then, the motion of the system (1.48) *is* x-*bounded uniformly in* y_0.

The proof of Theorem 1.10 is similar to that of Theorem 1.8.

1.4.4 Applications

1.4.4.1 *General case of a system with* fast and slow variables

Consider a system of perturbed motion equations containing fast and slow variables (see [23, 46] and the bibliography therein):

$$\begin{aligned}
\frac{dx}{dt} &= \mu X(t, x, y, \mu), \\
\frac{dy}{dt} &= Y(t, x, y) + \mu F(t, x, y, \mu),
\end{aligned} \tag{1.64}$$

where $x \in \mathbb{R}^n$, $y \in \mathbb{R}^m$, $X \in C(\mathbb{R}_+ \times \mathbb{R}^n \times \mathbb{R}^n \times M, \mathbb{R}^n)$, $Y \in C(\mathbb{R}_+ \times \mathbb{R}^n \times \mathbb{R}^m, \mathbb{R}^m)$, $F \in C(\mathbb{R}_+ \times \mathbb{R}^n \times \mathbb{R}^m \times M, \mathbb{R}^m)$, and $M = (0, 1]$ is the range of a small parameter μ. Setting $\mu = 0$ in the system (1.64), we get a *degenerate system*:

$$\begin{aligned}
\frac{dx}{dt} &= 0, \quad x(t_0) = x_0, \\
\frac{dy}{dt} &= Y(t, x, y), \quad y(t_0) = y_0.
\end{aligned} \tag{1.65}$$

If for the vector function $Y(t, x, y)$, there exists a continuous non-negative function $\bar{\psi}_2(t)$ such that the inequality

$$\|Y(t, x, y)\| \leq \bar{\psi}_2(t)(\|x\| + \|y\|)^q, \quad q > 1,$$

is true, then according to Corollary 1.4, we find the estimate

$$\|z(t)\| \leq \|z_0\|(1 - N_2^*(t_0, t))^{-\frac{1}{q-1}}$$

for all $t \in J \subseteq \mathbb{R}_+$, for which

$$N_2^*(t_0, t) = (q-1)\|z_0\|^{q-1} \int_{t_0}^{t} \bar{\psi}_2(s)ds < 1.$$

Suppose that for the right-hand side of the system (1.64), there are continuous and non-negative functions $\varphi_i(t, \mu)$, $i = 1, 2$, and the following inequalities:

A_7.

$$\|Y(t,x,y,\mu)\| + \|F(t,x,y,\mu)\| \le \varphi_2(t,\mu)(\|x\| + \|y\|)^q, \quad q > 1$$

for $0 < \mu < \mu_1$, and

A_8.

$$\|X(t,x,y,\mu)\| \le \varphi_1(t,\mu)(\|x\| + \|y\|)^p, \quad p > 1$$

for $0 < \mu < \mu_2$. Then, it follows from the equations (1.64) that

$$\|z(t)\| \le \|z_0\| + \int_{t_0}^t \varphi_1(s,\mu)\|z(s)\|^p ds$$

$$+ \int_{t_0}^t \varphi_2(s,\mu)\|z(s)\|^q ds, \quad t \ge t_0. \tag{1.66}$$

Applying Lemma 1.7 to the inequality (1.66), we get the estimate

$$\|z(t)\| \le \|z_0\| \left(1 - N(t_0,t,\mu)\right)^{-\frac{1}{p+q-2}} \tag{1.67}$$

for all $t \in J \subseteq \mathbb{R}_+$, for which

$$N(t_0,t,\mu) = (p+q-2)\Bigg[\|z_0\|^{p-1} \int_{t_0}^t \varphi_1(s,\mu)ds$$

$$+ \|z_0\|^{q-1} \int_{t_0}^t \varphi_2(s,\mu)ds\Bigg] < 1$$

for $0 < \mu < \min(\mu_1, \mu_2)$.

Following the idea of the proof of Theorem 1.7, it is easy to show that the motion of the system (1.58) is x-bounded if the following conditions are satisfied:

(1) $N(t_0,t,\mu) < 1$ for all $t \in J$ and $0 < \mu < \mu_3$;
(2) for some $\lambda^* > 0$, for a given value of $\beta^* > 0$, the following inequality is true:

$$\frac{1}{\lambda^*}\left(1 - N(t_0,t,\mu)\right)^{-\frac{1}{p+q-2}} < \frac{\beta^*}{\|z_0\|}$$

for all $0 < \mu < \min(\mu_1, \mu_2, \mu_3)$.

1.4.4.2   *Reduction of the system (1.64) to the standard form*

Let the solution of the system (1.65) be

$$x = const \quad \text{and} \quad y = \varphi(t,x,c), \tag{1.68}$$

where $\varphi \in C(\mathbb{R}_+ \times \mathbb{R}^n \times \mathbb{R}, \mathbb{R}^m)$ and c is an arbitrary constant. Considering (1.68) as a change of variables in the system (1.64), it is easy to obtain the following system of *equations in standard form*:

$$\frac{dx}{dt} = \mu X(t, x, \varphi(t, x, c), \mu), \quad x(t_0) = x_0,$$
$$\frac{dc}{dt} = \mu C(t, x, c, \mu), \quad c(t_0) = c_0,$$

(1.69)

where

$$C(t, x, c, \mu) = \mu \left(\frac{\partial \varphi}{\partial c}\right)^{-1} \left\{ F(t, x, \varphi(t, x, c), \mu) - \frac{\partial \varphi}{\partial x} X(t, x, \varphi, \mu) \right\}$$

and c_0 is determined from the equation

$$y_0 = \varphi(t_0, x_0, c_0).$$

(1.70)

For the system (1.64), we introduce the following assumptions:

A_9. The functions $X(t, x, y, \mu)$, $Y(t, x, y)$, and $F(t, x, y, \mu)$ are defined and continuous in the domain

$$Q = \{t \geq t_0, \ x \in D_1, \ y \in D_2, \ 0 < \mu \leq \mu_0\}.$$

A_{10}. The solution $y = \varphi(t, x, c)$ is defined in the domain $Q_1 = \{t \geq t_0,$ $x \in D_1, \ c \in D_3\}$, and $\varphi(t, x, c) \in D_2$ as soon as $(t, x, c) \in Q_1$ and the partial derivatives $\dfrac{\partial \varphi_i}{\partial x_j}$ and $\dfrac{\partial \varphi_i}{\partial c_j}$ are bounded in Q_1.

A_{11}. There are continuous non-negative functions $\psi_1(t, \mu)$ and $\psi_2(t, \mu)$, and the following inequalities hold in $Q \times Q_1$:

(a) $\|X(t, x, \varphi(t, x, c), \mu)\| \leq \psi_1(t, \mu)(\|x\| + \|c\|)^p,$

(b) $\|C(t, x, c, \mu)\| \leq \psi_2(t, \mu)(\|x\| + \|c\|)^q,$

(1.71)

where $p > 1$, $q \geq 1$, and $\mu \in (0, \mu_0)$.

If conditions A_9–A_{11} are satisfied, Lemma 1.7 is applicable to the system (1.69), which leads to the following estimate of solutions:

$$\|\omega(t)\| \leq \|\omega_0\| (1 - \Phi(t_0, t, \mu))^{-\frac{1}{p+q-2}}$$

for all $t \in J$, for which

$$\Phi(t_0, t, \mu) = (p + q - 2)\left[\|\omega_0\|^{p-1} \int_{t_0}^t \psi_1(s, \mu)ds \right.$$
$$\left. + \|\omega_0\|^{q-1} \int_{t_0}^t \psi_2(s, \mu)ds \right] < 1$$

and $\mu \in (0, \mu_0)$.

Here, $\|\omega(t)\| = \|x(t)\| + \|c(t)\|$ for all $t \in J$ and $\|\omega_0\| = \|x_0\| + \|c_0\|$.

1.4.4.3   *Systems with several rotating phases*

These types of systems include equations of the form

$$\frac{dx}{dt} = \mu X(t, x, y),$$
$$\frac{dy}{dt} = \omega(x) + \mu Y(t, x, y), \tag{1.72}$$

where $x \in \mathbb{R}^n$, $y \in \mathbb{R}^m$ is the phase vector $\omega : \mathbb{R}^n \to \mathbb{R}^m$ and the vector functions $X(t, x, y)$ and $Y(t, x, y)$ are periodic in y with period 2π.

Let, in the system (1.72), $\mu = 0$. Then,

$$x = const, \; y = \omega(x)t + c \tag{1.73}$$

are solutions to the degenerate system corresponding to (1.72).

As in Section 1.3.4.2, the system (1.72) is reduced to the standard form

$$\frac{dx}{dt} = \mu X(t, x, \omega(x)t + c), \; x(t_0) = x_0,$$
$$\frac{dy}{dt} = \mu Y(t, x, \omega(x)t + c), \; y(t_0) = y_0. \tag{1.74}$$

Suppose that for any non-zero integer vector k $(\omega(x), k) \neq 0$ and

$$|(\omega(x), k)| > \frac{const}{|k|^{m+1}}, \quad |k| = |k_1| + \cdots + |k_m|,$$

for any $x \in D_1 \subset \mathbb{R}^n$, i.e., there is no resonance in the system (1.74) for all $t \geq t_0$.

Let us make certain assumptions about the system (1.74):

A_{12}. There are continuous non-negative functions $a(t, \mu)$, $b(t, \mu)$ such that

(a) $\mu\|X(t, x, \omega(x)t + c)\| \leq a(t, \mu)(\|x\| + \|c\|)^p$
for all $(t, x, c) \in \mathbb{R}_+ \times \mathbb{R}^n \times \mathbb{R}^m$, $\mu \in (0, \mu_0)$;

(b) $\mu\|Y(t, x, \omega(x)t + c)\| \leq b(t, \mu)(\|x\| + \|c\|)^q$
for all $(t, x, c) \in \mathbb{R}_+ \times \mathbb{R}^n \times \mathbb{R}^m$ and some $p \geq 1$, $q > 1$.

Under the conditions of assumption A_{12}, Lemma 1.7 is applicable to the system (1.74), which makes it possible to establish an estimate for $\|z(t)\| = \|x(t)\| + \|c(t)\|$ in the form

$$\|z(t)\| \leq \|z_0\|(1 - \Pi(t_0, t, \mu))^{-\frac{1}{p+q-2}} \tag{1.75}$$

for all $t \geq t_0$, for which

$$\Pi(t_0, t, \mu) = (p + q - 2) \left[\|z_0\|^{p-1} \int_{t_0}^{t} a(s, \mu)ds + \|z_0\|^{q-1} \int_{t_0}^{t} b(s, \mu)ds \right] < 1$$

for $\mu \in (0, \mu_0)$.

The estimate (1.75) allows us to establish conditions for the x-boundedness of the motion of the system (1.72) in the same way as we did in Section 1.3.3.

1.5   Infinite Systems of Equations

In this section, for *infinite systems* of ordinary differential equations (ISODEs), new bounds for the norm of solutions are established and their application is shown for estimating the deviation of solutions of averaged equations from exact solutions of standard ISODEs.

1.5.1   Notation and definitions

Let us recall some results necessary for further presentation.

Let W be the set of all complex sequences $x = (x_k)_{k=1}^{\infty}$. For any value $1 \leq p < \infty$, we denote $l_p = \{x \in W : \sum_{k=1}^{\infty} |x_k|^p < \infty\}$ and consider a *Banach space* with the norm $\|x\|_p = (\sum_{k=1}^{\infty} |x_k|^p)^{\frac{1}{p}}$.

The space l_p is metric with the norm

$$\rho(x, y) = \left(\sum_{k=1}^{\infty} |x_k - y_k|^p \right)^{\frac{1}{p}}, \quad 1 \leq p < \infty,$$

where $(x_k) \in l_p$, $(y_k) \in l_p$ for all $k = 1, 2, \ldots, \infty$.

The scalar product of the vectors $x \in l_p$ and $y \in l_p$ is calculated using the formulas $(x, y) = \sum_{k=1}^{\infty} x_k y_k$ and $|(x, y)| \leq \|x\|_p \|y\|_p$.

Let S and M be the subsets of the *metric space* (l_p, ρ) and $\varepsilon > 0$. A subset S is called a ε-network of a subset M in l_p if for any $x \in M$, there exists an element $s \in S$ such that $\rho(x, s) < \varepsilon$.

If the set S is finite, then the ε-network is called a *finite ε-network*.

Let M_{l_p} be the set of all bounded subsets in the metric space (l_p, ρ).

If $A \in M_{l_p}$, then the *measure of non-compactness* of the Kuratowski (see [64]) set A is defined as the infimum of the set of all real $\varepsilon > 0$ such that A can be covered by a finite number of spheres of radius $r < \varepsilon$ centred at l_p. It is expressed as

$$\alpha(A) = \inf\{\varepsilon > 0 : A \text{ has a finite } \varepsilon\text{-net}\}.$$

The function $\alpha : M_{l_p} \to [0, \infty)$ is called the *Kuratowski measure* of non-compactness.

Let A, A_1, and A_2 be bounded subsets in the metric space (l_p, ρ). Then, the following relations are true:

(i) $\alpha(A) = 0$ if and only if $\overline{A}$ is compact, where $\overline{A}$ denotes the closure of A;

(ii) if $A_1 \subset A_2$, then $\alpha(A_1) \leq \alpha(A_2)$;

(iii) $\alpha(A_1 + A_2) \leq \alpha(A_1) + \alpha(A_2)$;

(iv) if $A_1 = \{x_n\}$, $x_n \in l_p$, $A_2 = \{y_n\}$, $y_n \in l_p$, then

$$\alpha(\{x_n\}) - \alpha(\{y_n\}) \leq \alpha(\{x_n - y_n\});$$

(v) $\alpha(A) = \alpha(co(A))$, where $co(A)$ is the *convex hull of A*.

Other relations for the measure of non-compactness $\alpha : M_{l_p} \to [0, \infty)$ and their application in the theory of equations are given in the books [13, 64, 68].

1.5.2   Statement of the problem

The ISODE of perturbed motion is considered in the form

$$\frac{dx}{dt} = f(t, x_1, x_2, \ldots), \tag{1.76}$$

$$x(t_0) = x_0, \tag{1.77}$$

where $x \in \mathbb{R}^\infty$, $t \in J = [t_0, t_0 + a]$, $a > 0$, $f \in C(D, l_p)$, $D = \{(t, x) \in J \times B[x_0, b]\}$, and $B[x_0, b] = \{x \in l_p : \|x - x_0\|_p < b\}$, $b > 0 - const.$

The motion of some mechanical or any other natural system is well-defined if the solution of the Cauchy problem (1.76)–(1.77) exists on the interval J and $x(t) \in l_p$.

Together with the problem (1.76)–(1.77), we consider the associated scalar initial value problem

$$\frac{du}{dt} = g(t, u), \tag{1.78}$$

$$u(t_0) = u_0 \geq 0, \tag{1.79}$$

where $u \in \mathbb{R}$ and $g \in C(J \times [0, 2b], \mathbb{R})$.

We indicate the conditions for the local existence of solutions to the initial problem (1.76)–(1.77) and establish an estimate for the norm of solutions to the system (1.76).

1.5.3   Main result

The solution of the problem posed consists of two parts: the first is obtaining conditions for the local existence of solutions to the problem (1.76)–(1.77), and the second is establishing an *estimate* for the *norm of solutions* to the problem (1.76)–(1.77) under broad assumptions about the right-hand side of the system (1.76).

To this end, we introduce some assumptions:

A_{13}. Initial conditions for problem (1.76)–(1.77) are $x_0 \in l_p$.

A_{14}. There is a constant $M > 0$ such that $\|f(t, x)\|_p \leq M$ on $D_0 \subseteq D$ and

$$J_1 = [t_0, t_0 + \tau], \tau = \min\left\{a, \frac{b}{M+1}\right\}.$$

A_{15}. The vector function $f \in C(D_0, l_p)$ and is uniformly continuous on D.

A_{16}. For any set $A \subset B[x_0, b]$, the Kuratowski *measure of non-compactness* of the right-hand side of the equation (1.76) satisfies the inequality

$$\alpha(f(t, A)) \leq g(t, \alpha(A)). \tag{1.80}$$

A_{17}. $g \in C(J \times [0, 2b], \mathbb{R})$, $g(t, 0) = 0$, and $u(t) \equiv 0$ is the only solution to the problem (1.78)–(1.79).

A_{18}. There are constant $m > 1$ and continuous non-negative functions $a(t)$ and $b(t)$ such that

$$\|f(t, x)\|_p \leq a(t)\|x\|_p + b(t)\|x\|_p^m, \quad m > 1,$$

for all $(t, x) \in D_0$.

The following assertion holds.

Theorem 1.11. *Let the conditions of the assumptions A_{13}–A_{18} be satisfied for the ISODE (1.76) and, moreover,*

$$M(t_0, t) = (m-1)\|x_0\|_p^{m-1} \int_0^t b(s) \exp\left[(m-1)\int_0^s a(\tau)d\tau\right] ds < 1 \tag{1.81}$$

for all $t \in J_2 \subset J$. Then, the solution $x(t)$ of the ISODE (1.76) exists on J_1, and the norm of the solutions satisfies the estimate

$$\|x(t)\|_p \leq \|x_0\|_p \exp\left(\int_0^t a(s)ds\right)(1 - M(t_0, t))^{-\frac{1}{m-1}} \tag{1.82}$$

for all $t \in J_1 \cap J_2$.

Proof.   Under conditions A_{13}–A_{18} for the problem (1.76)–(1.77), there exists an *approximate solution* $\{x_n(t)\}$ on $[t_0, t_0 + \tau]$ such that

$$\frac{dx_n(t)}{dt} = f(t, x_{1n}(t), x_{2n}(t), \ldots) + y_n(t), \qquad (1.83)$$

$$x_n(t_0) = x_0 \in l_p \qquad (1.84)$$

and $\|y_n(t)\|_p \leq \varepsilon_n$, where $\varepsilon_n \to 0$ as $n \to \infty$. $\square$

The sequence $\{x_n(t)\}$ is equi-continuous and uniformly bounded; therefore, for the existence of solutions to the problem (1.76)–(1.77), it suffices to show that the set $\{x_n(t)\} = \{x_n(t) : n \geq 1\}$ is relatively compact, i.e., the condition $\alpha(\{x_n(t)\}) \equiv 0$ is satisfied for all $t \in [t_0, t_0 + \tau]$.

In this case, the limit of the uniformly convergent sequence is the solution $x(t) \in l_p$ of the problem (1.76)–(1.77) for all $t \in [t_0, t_0 + \tau]$.

Further, we denote $B_k(t) = \{x_n(t) : n \geq k\}$ and consider the measure of non-compactness $m(t) = \alpha(B_k(t), m(t_0)) = 0$. Since $m(t)$ is continuous, due to property (ii) of the measure of non-compactness, we obtain

$$|m(t) - m(s)| \leq \alpha(\{x_n(t) - x_n(s) : n \geq k\}) \leq 2(M+1)|t - s|.$$

Let us show that

$$D_- m(t) \leq g(t, m(t)) \quad \text{for all} \ \ t \in [t_0, t_0 + \tau],$$

where

$$D_- m(t) = \liminf\{[m(t) - m(t - h)]h^{-1} : h \to 0^+\}.$$

Considering again property (ii) of the measure of non-compactness, we obtain

$$[m(t) - m(t - h)]h^{-1} \leq \alpha\{[x_n(t) - x_n(t - h)]h^{-1} : n \geq k\}.$$

Hence, by the mean value theorem (see [68], Theorem 1.5) and property (iii) of the non-compactness measure, we find that

$$D_- m(t) \leq \lim_{h \to 0^+} \inf \alpha(B_k^*(J_h)),$$

where $J_h = [t - h, t]$ and $B_k^*(J_h) = \bigcup_{t \in J_h} B_k^*(t)$, $B_k^*(t) = \{x_n^*(t) : n \geq k\}$.
In addition, for the solutions of the problem (1.76)–(1.77), we have an estimate of the measure of non-compactness:

$$\alpha(B_k^*(J_h)) \leq \alpha(f(J_h, B_k(J_h)) + \{y_n(J_h)\} \leq \alpha(f(J_h, B_k(J_h))) + 2\varepsilon_k.$$

The equi-boundedness of the sequence $\{x_n(t)\}$, the uniform continuity of f, and condition A_{13} of Theorem 1.11 guarantee the convergence

$$f(J_h, B_k(J_h)) \to f(t, B_k(t))$$

according to the Hausdorff measure. Hence, it follows that

$$D_- m(t) \leq g(t, m(t)) + 2\varepsilon_k, \quad t \in [t_0, t_0 + \tau]. \tag{1.85}$$

From (1.85) and the principle of comparison, it follows that

$$\alpha(\{x_n(t)\} : n \geq 1) = m(t) \leq r_k(t, t_0, 0), \quad t \in [t_0, t_0 + \tau],$$

where $r_k(t, t_0, 0)$ is the maximum solution of the initial problem

$$\frac{du}{dt} = g(t, u) + 2\varepsilon_k, \tag{1.86}$$

$$u(t_0) = 0. \tag{1.87}$$

From condition A_{17}, for the problem (1.86)–(1.87), we obtain $\lim r_k(t, t_0, 0) = r(t, t_0, 0)$, where $r(t, t_0, 0)$ is the maximum solution of problem (1.86)–(1.87). Therefore, $\alpha(\{x_n(t)\}) \equiv 0$ on $[t_0, t_0 + \tau]$, which proves the existence of a solution $x(t) \in l_p$ on $[t_0, t_0 + \tau]$.

Next, we show that the estimate (1.82) holds for the norm of solutions to the problem (1.76)–(1.77). For any $x(t) \in l_p$ and $t \in J$, under condition A_{18}, we have the estimate

$$\|x(t)\|_p \leq \|x_0\|_p + \int_0^t (a(s) + b(s)\|x(s)\|_p^{m-1})\|x(s)\|_p ds \tag{1.88}$$

for all $t \in J$. Hence, it follows that

$$\|x(t)\|_p \leq \|x_0\|_p \exp\left[\int_0^t (a(s) + b(s)\|x(s)\|_p^{m-1}) ds\right]. \tag{1.89}$$

Applying to the inequality (1.89) the same estimation technique as in the proof of Lemma 1.6, we obtain the estimate

$$\|x(t)\|_p^{m-1} \leq \|x_0\|_p^{m-1} \exp\left[(m-1)\int_0^t a(s) ds\right] (1 - M(t_0, t))^{-\frac{1}{m-1}},$$

$$t \in J_2. \tag{1.90}$$

Since $m > 1$, by the hypothesis of Theorem 1.11, (1.90) implies the estimate (1.82). Theorem 1.11 is proved.

Corollary 1.5. *Let conditions A_{13}–A_{17} be satisfied and, in condition A_{18}, the function $b(t) \equiv 0$ for all $t \in J$. Then, the inequality (1.90) takes the form*

$$\|x(t)\|_p \leq \|x_0\|_p + \int_0^t a(s)\|x(s)\|_p ds, \quad t \in J,$$

and according to the Gronwall–Bellman lemma, we get

$$\|x(t)\|_p \leq \|x_0\|_p \exp\left(\int_0^t a(s)ds\right)$$

for all $t \in J$ and $\|x_0\|_p > 0$.

Corollary 1.6. *Let conditions A_{13}–A_{17} be satisfied and, in condition A_{18}, the function $a(t) \equiv 0$ for all $t \in J$. Then, the inequality (1.90) takes the form*

$$\|x(t)\|_p \leq \|x_0\|_p + \int_0^t b(s)\|x(s)\|_p^m ds, \quad t \in J.$$

It is easy to show that

$$\|x(t)\|_p \leq \|x_0\|_p (1 - M_1(t_0, t))^{-\frac{1}{m-1}}$$

for all $t > 0$, for which

$$M_1(t_0, t) = (m - 1)\|x_0\|_p^{m-1} \int_0^t b(s)ds < 1$$

for any $x_0 \in l_p$, $x_0 > 0$.

1.5.4   Quasi-linear systems

An important special case of the ISODE (1.76) is the *quasi-linear counting system* [136, 150]

$$\frac{dx_s}{dt} = \sum_{k=1}^{\infty} p_{sk}(t)x_k + N_s(t, x_1, x_2, \ldots), \quad s = 1, 2, \ldots, \tag{1.91}$$

where $p_{sk}(t) : \mathbb{R}_+ \to \mathbb{R}$ and $N_s \in C(\mathbb{R}_+ \times l_p, l_p)$ for all $s = 1, 2, \ldots$.

Let us establish an estimate for the norm of the solutions of the ISODE (1.91) under the initial conditions (1.77).

Denote $P(t) = \sum_{k=1}^{\infty} p_{sk}(t)$, $s = 1, 2, \ldots$, and $N(t, x) = (N_1(t, x_1, x_2, \ldots), N_2(t, x_1, x_2, \ldots), \ldots)^T$.

Let us assume that the conditions for the existence of solutions are satisfied and assumption A_{18}^* holds. On J, there exist continuous non-negative functions $\bar{a}(t), \bar{b}(t)$ such that

$$\|P(t)\|_p \leq \bar{a}(t) \text{ for all } t \in J \text{ and } \|N(t,x)\|_p \leq \bar{b}(t)\|x\|_p^2 \text{ for all } t \in J.$$

The following assertion holds.

Theorem 1.12. *Let the system* (1.91) *satisfy the assumption* A_{18}^* *and, in addition,*

$$M^*(t_0, t) = \|x_0\|_p \int_{t_0}^t \bar{a}(s) \left(\exp \int_0^s \bar{b}(\tau) d\tau \right) ds < 1$$

for all $t \in J_2$.

Then, a solution $x(t)$ *of the system* (1.91) *exists on* J_1, *and the norm of the solutions satisfies the estimate*

$$\|x(t)\|_p \leq \|x_0\|_p \exp \left(\int_{t_0}^t \bar{a}(s) ds \right) (1 - M^*(t_0, t))^{-1}$$

for all $t \in J_1 \cap J_2$.

Proof. For $x(t) \in l_p$ from (1.91), we have

$$x(t) = x_0 + \int_{t_0}^t \left(P(s)x(s) + N(s, x(s)) \right) ds.$$

Hence, under conditions A_{16}^*, it follows that

$$\|x(t)\|_p \leq \|x_0\|_p + \int_{t_0}^t \left(\bar{a}(s) + \bar{b}(s)\|x(s)\|_p \right) \|x(s)\|_p ds,$$

and moreover,

$$\|x(t)\|_p \leq \|x_0\|_p \exp \left(\int_{t_0}^t (\bar{a}(s) + \bar{b}(s)\|x(s)\|_p) \, ds \right),$$

or

$$-\|x(t)\|_p \exp \int_{t_0}^t \left(-\bar{b}(s)\|x(s)\|_p \right) ds \geq -\|x_0\|_p \exp \left(\int_{t_0}^t \bar{a}(s) ds \right).$$

Multiplying both sides of this inequality by $\bar{b}(t)$, we obtain the inequality

$$\frac{d}{dt} \left\{ \exp \left(-\int_{t_0}^t \bar{b}(s)\|x(s)\|_p ds \right) \right\} \geq -\|x_0\|_p \bar{b}(t) \exp \left(\int_{t_0}^t \bar{a}(s) ds \right),$$

which upon integrating between t_0 and t, we find

$$(\|x(t)\|_p)^{-1}\|x_0\|_p$$

$$\times \exp\left(\int_{t_0}^t \bar{a}(s)ds\right) \geq 1 - \|x_0\|_p \int_{t_0}^t \bar{b}(s)\exp\left(\int_{t_0}^t \bar{a}(\tau)d\tau\right)ds. \quad \square$$

Considering that by the condition of Theorem 1.12, the expression $M^*(t_0,t) < 1$ holds for all $t \in J_2$, we obtain the estimate $\|x(t)\|_p$ for all $t \in J_1 \cap J_2$. Theorem 1.12 is proved.

1.5.5   Applications

Consider an ISODE in the standard form [174]:

$$\frac{dx_k}{dt} = \mu F_k(t,\tau,x_1,x_2,\ldots), \tag{1.92}$$

$$x_k(t_0) = x_k^0 \in l_p, \tag{1.93}$$

where $\tau = \mu t$, $\mu > 0$, and $(x_k) \in l_p$, $F_k : \mathbb{R}_+^2 \times \mathbb{R}^\infty \to \mathbb{R}$ are continuous mappings $\mathbb{R}_+^2 \times l_p \to l_p$, $k = 1, 2, \ldots$.

Suppose that in the region $\|x\|_p \leq a$, $0 \leq \tau \leq L$, where $a > 0$ and $L > 0$ are *const*, there exists an average

$$F_{k0}(\tau,x_1,x_2,\ldots) = \lim\left\{\left(\int_0^T F_k(t,\tau,x_1,x_2,\ldots)dt\right)T^{-1} : T \to \infty\right\},$$

where integration is performed only over t.

Along with the Cauchy problem for (1.92)–(1.93), we consider the initial problem for the averaged ISODE

$$\frac{dy_k}{dt} = \mu F_{k0}(\tau,y_1,y_2,\ldots), \tag{1.94}$$

$$y_k(t_0) = y_k^0 \in l_p. \tag{1.95}$$

Let for $0 < \mu < \mu_0$, $\mu_0 \in (0,1]$, for the problems (1.92)–(1.93) and (1.94)–(1.95), the conditions similar to those given in assumptions A_1–A_5 be satisfied and, in addition, the following assumption hold:

A_{19}. There are functions $a(t,\mu)$ and $b(t,\mu)$ continuous and non-negative for all $t \in [0,T]$ such that

$$\|F(t,\mu t,x) - F_{k0}(\mu t,y)\|_p \leq a(t,\mu)\|x-y\|_p + b(t,\mu)\|x-y\|_p^m,$$
$$m > 1. \tag{1.96}$$

From the equations (1.92) and (1.94), under the initial conditions (1.93) and (1.95), we obtain

$$x(t) - y(t) = x_0 - y_0 + \mu \int_0^t [F(s, \mu s, x(s)) - F_{ko}(\mu s, y(s))]ds$$

and, further,

$$\|x(t) - y(t)\|_p \le \|x_0 - y_0\|_p + \int_0^t (a(s,\mu)\|x(s) - y(s)\|_p + b(s,\mu)\|x(s)$$
$$-y(s)\|_p^m)ds, t \ge 0. \tag{1.97}$$

Applying Theorem 1.12 to the inequality (1.97), we get the assertion. Let $0 < \|x_0 - y_0\|_p < \infty$ and

$$M(t_0, t, \mu) = (m-1)\|x_0 - y_0\|_p^{m-1} \int_0^t b(s,\mu) \times \exp\left[(m-1)\int_0^s a(\tau,\mu)d\tau\right] ds < 1$$

for all $0 \le \mu t \le L$. Then,

$$\|x(t) - y(t)\|_p \le \|x_0 - y_0\|_p \exp\left(\int_0^t a(s,\mu)ds\right)\left(1 - M(t_0,t,\mu)\right)^{-\frac{1}{m-1}} \tag{1.98}$$

for all $0 \le \mu t \le L$.

It follows from the bound (1.98) that for any $\varepsilon > 0$, the deviation of solutions of the averaged problem (1.94)–(1.95) from those of the original (1.92)–(1.93) satisfies the estimate

$$\|x(t) - y(t)\|_p < \varepsilon \tag{1.99}$$

if the values $\delta > 0$ and $\mu_0 > 0$ are chosen so that for $\|x_0 - y_0\|_p < \delta$ and $\mu < \mu_0$, the inequality

$$\exp\left(\int_0^t a(s,\mu)ds\right)\left(1 - M(t_0,t,\mu)\right)^{-\frac{1}{m-1}} < \frac{\varepsilon}{\delta} \tag{1.100}$$

holds for all $0 < \mu t < L$.

Consider some special cases of the condition (1.96).

Case 1. Let the bound (1.96) be a function in $b(t,\mu) \equiv 0$ for all $0 < \mu t < L$. Then, the inequality (1.97) takes the form

$$\|x(t) - y(t)\|_p \le \|x_0 - y_0\|_p + \int_0^t a(s,\mu)\|x(s) - y(s)\|_p ds,$$

from which it follows that

$$\|x(t) - y(t)\|_p \le \|x_0 - y_0\|_p \exp\left(\int_0^t a(s,\mu)ds\right) \qquad (1.101)$$

for all $0 < \mu t < L$. From the estimate (1.101), we find that $\|x(t) - y(t)\|_p < \varepsilon$ if

$$\exp\left(\int_0^t a(s,\mu)ds\right) < \frac{\varepsilon}{\delta} \qquad (1.102)$$

for all $0 \le \mu t \le L$ as soon as $0 < \|x_0 - y_0\| < \delta$ and $0 < \mu < \mu_0$. The inequality (1.102) will be true if

$$\int_0^t a(s,\mu)ds < \ln\left(\frac{\varepsilon}{\delta}\right) \quad \text{for all } 0 \le \mu t \le L.$$

Case 2. Let the estimate (1.96) be the function $a(t,\mu) \equiv 0$ for all $0 \le \mu t \le L$. Then, the inequality (1.97) takes the form

$$\|x(t) - y(t)\|_p \le \|x_0 - y_0\|_p + \int_0^t b(s,\mu)\|x(s) - y(s)\|_p^m ds. \qquad (1.103)$$

We introduce the notation

$$M_1(t_0, t, \mu) = (m-1)\|x_0 - y_0\|_p^{m-1}\int_0^t b(s,\mu)ds, \qquad (1.104)$$

and we assume that $M(t_0, t, \mu) < 1$ for all $0 \le \mu t \le L$. Applying Corollary 1.6 to the inequality (1.103), we get the estimate

$$\|x(t) - y(t)\|_p \le \|x_0 - y_0\|_p^{m-1}\left(1 - M_1(t_0, t, \mu)\right)^{-\frac{1}{m-1}} \qquad (1.105)$$

for all $0 \le \mu t \le L$.

It follows from the bound (1.97) that $\|x(t) - y(t)\|_p < \varepsilon$ for all $0 \le \mu t \le L$ if for $\|x_0 - y_0\| < \delta$ and $\mu < \mu_0$, the inequalities $M_1(t_0, t, \mu) < 1$ hold and

$$\left(1 - M_1(t_0, t, \mu)\right)^{-\frac{1}{m-1}} < \frac{\varepsilon}{\delta} \quad \text{for all } 0 \le \mu t \le L.$$

1.6   Comments and Bibliography

Linear integral inequalities are widely used in the theory of differential equations, starting with the Gronwall–Bellman inequality (see Bellman [14, 15], Aleksandrov [3], Bainov and Simeonov [12], Gronwall [47], Chezary [35], Hubbard and West [54], Lungu and Ciplea [75], Oguntuase [128], Rozou [141], etc.).

The results of the development of the theory of nonlinear integral inequalities are presented in many papers and monographs, including Alami [2], Beesack [18], Bihari [19], Gutowski and Radziszewski [45], Lakshmikantham, Leela, and Martynyuk [66], and N'Doye, Zasadzinski, Darouach, Radhy, and Bouaziz [125], Pachpatte [129].

Many works are devoted to estimating solution norms, for example, Brauer [24], Coddington and Levinson [31], Langenhop [69], and Martynyuk [82, 83]. Classical results of the analysis of the boundedness of solutions of differential equations are given in the monograph by Yoshizawa [169] and many papers (see Raffoul [143] and bibliography there).

This chapter presents estimates of the norms of solutions of nonlinear systems of differential equations obtained on the basis of nonlinear integral inequalities.

The results of Section 1.2 are new.

Section 1.3 is based on the papers by Martynyuk [83] and Martynyuk, Khusainov, and Chernienko [107]. Quasi-linear systems are widely used in modelling real-world phenomena; see, for example, Fieguth [40], Frangos [41], Hayashi [53], Krasovsky [61], Malkin [77], Martynyuk *et al.* [100].

The results of Section 1.4 are new.

In many problems of continuum mechanics, including oscillations of systems with distributed parameters, dynamic stability of elastic systems, and a number of others, partial differential equations are approximated by ISODEs (see Martynyuk [85] and the bibliography therein). The study of infinite systems of equations originates from the works of Hale [51], Hart [52], Mursalem [118], Reid [144], and Roxin [146]. In the work of Tikhonov [164], a fundamental result was obtained on the unique solvability of the Cauchy problem for ISODEs. Lyapunov's stability theory (see Lyapunov [76]) was extended to ISODEs in the works of Persidski (see Persidski [136] and the bibliography therein). The method of averaging in nonlinear mechanics for ISODEs was considered in the work of Zhautukov [174].

As for ordinary differential equations of perturbed motion for ISODEs, the problem of estimating the norm of solutions remains relevant since its solution allows us to establish new conditions for the stability of motion, estimates of approximate integrations, etc.

Estimates for the norms of solutions to ISODEs are given in many papers based on the Gronwall–Bellman lemma, conditioned by the linear estimate of nonlinearities; see Samoilenko and Teplinski [150].

The boundaries for solutions obtained in Section 1.5 allow us to more accurately take into account the influence of the nonlinear terms of the

ISODE on the dynamics of the entire system and are therefore of interest for applications. These estimates make it possible to establish sufficient conditions for boundedness, stability on a finite interval, and a number of other properties of solutions to the ISODE.

The basis of integral methods of analysis in the qualitative theory of equations are integral equations [116], linear and nonlinear integral inequalities [80], homogenisation procedures, duality methods, optimal design, and conformal methods (see [33] and the bibliography therein). In the monograph [43], a numerical method for the optimal choice of parameters in the dynamics of mechanical systems is developed.

The stability and constraint criteria presented in this chapter have significant potential for application in aerospace mechanics (see Skullestad and Gilbert [158], Slemrod [159], Zou An-Min [175]).

Chapter 2

Integral Estimates of Lyapunov Functions and Their Applications

2.1 Introduction

In this chapter, we consider systems of equations of perturbed motion, for which we establish new estimates for the Lyapunov functions and indicate new conditions for various types of boundedness and stability of motion.

The chapter is organised according to the following plan.

Section 2.2 presents the formulation of the problem of estimating the Lyapunov function on solutions of nonlinear equations of perturbed motion and their connection with the comparison principle.

Section 2.3 considers a system of equations with a quadratic nonlinearity and establishes an estimate for the Lyapunov function for the case when its total derivative satisfies a quadratic differential inequality due to the equations of motion. For the basis of the obtained estimate, the conditions of boundedness, uniform boundedness, and equi-boundedness of motion are established, and the boundedness of motion with respect to some variables is investigated. In addition, conditions for boundedness of motion with respect to given areas of initial and subsequent perturbations are obtained here.

In Section 2.4, the general case of a quasi-linear system is considered. Here, we establish a new estimate for the Lyapunov function on solutions of the quasi-linear system and obtain stability conditions on a finite interval.

In Section 2.5, for nonlinear systems with an asymptotic expansion of the right-hand side, new estimates for the Lyapunov function are obtained. Based on these estimates, the conditions for the closeness of the solutions of the original and approximate systems of equations are established.

In Section 2.6, the results of dynamic analysis of solutions of essentially nonlinear systems are given.

The final section provides a brief overview of the development of the Lyapunov function method and bibliography related to this chapter.

2.2　Preliminary Analysis

The following nonlinear system of equations of perturbed motion is considered:

$$\frac{dx}{dt} = f(t, x) \tag{2.1}$$

under the initial conditions

$$x(t_0) = x_0, \tag{2.2}$$

where $x \in \mathbb{R}^n$, $f \in C(\mathbb{R}_- \times \mathbb{R}^n, \mathbb{R}^n)$, and $f(t, 0) = 0$ for all $t \geq t_0$, $\mathbb{R}_+ = [0, \infty)$. It is assumed that under the initial conditions $(t_0, x_0) \in (\mathbb{R}_+ \times \mathbb{R}^n)$, the solution $x(t, t_0, x_0)$ of the system (2.1)–(2.2) exists for all $t \geq t_0$.

A. Poincaré's memoir [138] about the qualitative analysis of solutions to systems of differential equations (2.1) without their direct integration stimulated Lyapunov to create a solution analysis method based on auxiliary functions (see [76]).

Definition 2.1. The function $V \in C(\mathbb{R}_+ \times \mathbb{R}^n, \mathbb{R}_+), V(t, 0) = 0$ for all $t \geq t_0$ is called a *Lyapunov function* if it is differentiable, definitely positive, and decreasing for all $t \geq t_0$ in some neighbourhood of the origin of the phase space and, together with the *total derivative*

$$V'(t, x) = V_t(t, x) + (f(t, x),\ \nabla_x V(t, x))$$

by virtue of the system (2.1), resolves the problem of stability (instability) of the state $x = 0$ of the system (2.1).

The actual application of the direct Lyapunov method involves two stages: the first is the construction of an appropriate Lyapunov function, and the second is the estimation of the total derivative of the function due to the equations of perturbed motion. As a result, based on the general Lyapunov theorems and/or their generalisations, the result of the qualitative analysis of the properties of motion is obtained.

Recall that in Lyapunov's work [76], the functions V and their total derivatives were introduced in the following cases:

— in the proof of the stability theorem (see [76, p. 62]), the function $V(t, x)$ is sign-definite, $V'(t, x)$ is constant sign of opposite sign with $V(t, x)$, or is identically zero.

$$V'(t, x) \leq 0; \tag{2.3}$$

— when proving the instability theorem (see [76, p. 68]), the relation

$$V'(t, x) = \lambda V + W \tag{2.4}$$

is considered, where $\lambda > 0$ and W is either identically equal to zero or some function of constant sign;

— when considering the critical case of one root being equal to zero with the rest being negative (see [76, pp. 106, 109]), where the characteristic equation has two purely imaginary roots (see [76, pp. 128, 170]). In this case, the form of the functions V and their derivatives V' becomes much more complicated.

In [119], for the function V, estimates of the total derivative are of the form

$$V'(t, x) \leq f(V), \tag{2.5}$$

where $f(V) > 0$ for $0 < V \leq H, f(0) = 0$, and also

$$V'(t, x) \leq \varphi(t) f(V), \tag{2.6}$$

where $\varphi(t)$ is such that $\int_{t_0}^{t} \varphi(s) ds \leq M, M > 0$ is a *constant*, and

$$V'(t, x) \leq g_1(t) V + g_2(t) V^{1+k}, \quad k > 0, \tag{2.7}$$

where the functions $g_1(t)$ and $g_2(t)$ are positive and integrable for all $t \geq t_0$.

In fact, the estimates (2.3)–(2.7) were a prerequisite for the emergence of *comparison equations* of the form

$$\frac{du}{dt} = 0, \quad u(t_0) = u_0; \tag{2.8}$$

$$\frac{du}{dt} = \lambda u + w, \quad u(t_0) = u_0, \tag{2.9}$$

where $w \geq 0$;

$$\frac{du}{dt} = g(u), \quad u(t_0) = u_0; \tag{2.10}$$

$$\frac{du}{dt} = \varphi(t)g(u), \quad u(t_0) = u_0; \tag{2.11}$$

$$\frac{du}{dt} = g_1(t)u + g_2(t)u^{1+k}, \quad u(t_0) = u_0. \tag{2.12}$$

Conti's suggestion (see [32, 94]) to consider the estimate for $V'(t,x)$ in the form

$$V'(t,x) \leq \omega(t,V), \tag{2.13}$$

where ω is a real function defined on $\mathbb{R}_+ \times \mathbb{B}_r$, $\mathbb{B}_r = \{x \in \mathbb{R}^n : \| x \| < r\}$, was the final step in obtaining an estimate for the Lyapunov function in the form

$$V(t,x(t)) \leq r(t,t_0,u_0), \tag{2.14}$$

where $r(t,\cdot)$ is the maximum solution of the comparison equation

$$\frac{dr}{dt} = \omega(t,r(t)), \quad r(t_0) = r_0, \tag{2.15}$$

for all $t \in [t_0, T)$, where $T > t_0$ is right end of the existence of solutions to the system of equations (2.1) and the *comparison equation* (2.15).

Definition 2.2. A tuple consisting of the system (2.1), the function $V(t,x)$, its total derivative $V'(t,x)$, the majorising function $\omega(t,V)$, and the comparison equation (2.15) forms the basis of the *comparison principle* in the qualitative theory of equations if it allows one to obtain an estimate of the form (2.14) for all $t \in [t_0, T)$ for the condition $V(t_0,x_0) \leq r_0$.

The monograph [86] presents the main results obtained in the development of the comparison principle based on the scalar-, vector-, and matrix-valued Lyapunov functions and indicates some applications of this approach.

The generality of the evaluation of the function $V(t,x(t))$ in the form (2.14), in the absence of a solution in the general case of the comparison equation (2.15), stimulates the search for new estimates of the function $V(t,x)$, particularly by using nonlinear integral inequalities.

We present one of the possible approaches in this direction.

2.2.1 Lyapunov function estimate based on the inequality (2.6)

Along with the estimate (2.6), we consider the comparison equations

$$\frac{dr}{dt} = \varphi(t)g(r), \quad r(t_0) = r_0 \tag{2.16}$$

and

$$\frac{dq}{dt} = -\varphi(t)g(q), \quad q(t_0) = r_0. \tag{2.17}$$

Let us introduce the notation $J(r) = \int_0^r du/g(u)$. Brauer [24] showed that the solution to the equation (2.16) is the function

$$r(t) = J^{-1}\left(J(r_0) + \int_{t_0}^t \varphi(s)ds\right). \tag{2.18}$$

This solution exists for all t, for which $J(r_0) + \int_{t_0}^t \varphi(s)ds < R$, where $R = \int_0^\infty du/g(u)$.

Namely, if $r(t_0) = r_0$, then $t \in [0, T)$, where T is determined from the relation $\int_{t_0}^T \varphi(s)ds = \int_{r_0}^\infty du/g(u)$.

If $\int_{r_0}^\infty du/g(u) = \infty$, then $T = \infty$.

If $\int_{r_0}^\infty \varphi(s)ds < \int_{r_0}^\infty du/g(u) \leq \infty$, then $J(r_0) + \int_{t_0}^t \varphi(s)ds < R$ and hence $r(t) < \infty$, i.e., the solution $r(t)$ is bounded for all $0 \leq t < \infty$.

Similarly, for the equation (2.17), the solution of the comparison equation (2.17) is obtained in the form

$$q(t) = J^{-1}\left(J(r_0) - \int_{t_0}^t \varphi(s)ds\right), \tag{2.19}$$

which is defined for all $t > t_0$, for which $J(r_0) - \int_{t_0}^t \varphi(s)ds > 0$.

Hence, the solution $q(t)$ exists for all $t \in [t_0, \tau)$, where τ is determined from the relation $\int_{t_0}^\tau \varphi(s)ds = \int_0^{r_0} du/g(u)$. If $\int_{t_0}^t \varphi(s)ds \leq \int_0^{r_0} du/g(u)$, then $q(t)$ exists for all $t_0 \leq t < \infty$ and if $\int_{t_0}^\infty \varphi(s)ds < \int_0^{r_0} du/g(u)$, then $q(t) > 0$ for all $t_0 \leq t < \infty$.

Hence, on the basis of the results from the paper [24], we obtain the following assertions.

Theorem 2.1. *If the condition (2.6) is satisfied for $0 < V \leq H$, $f(0) = 0$, and $f(V) > 0$, then*

$$V(t, x(t)) \leq J^{-1}(J(r_0) + \int_{t_0}^t \varphi(s)ds) \tag{2.20}$$

for $V(t_0, x_0) \le r_0$ for all $t \in [t_0, T)$, where T is determined from the relation

$$\int_{t_0}^{T} \varphi(s)ds = \int_{0}^{\infty} du/g(u). \qquad (2.21)$$

Theorem 2.2. *If the condition (2.6) is satisfied for $0 < V \le H$, $f(0) = 0$, and $f(V) > 0$, then*

$$V(t, x(t)) \ge J^{-1}\left(J(r_0) - \int_{t_0}^{t} \varphi(s)ds\right)$$

for $V(t_0, x_0) \ge r_0$ and for all $t \ge t_0$, for which $\int_{t_0}^{t} \varphi(s)ds \le \int_{0}^{\infty} du/g(u)$.

Theorem 2.3. *If the inequality $V' \le \varphi(t)f(V)$ holds, then the function $V(t, x(t))$ with the initial condition $V(t_0, x_0) \le r_0$ is defined on the interval $[0, T)$, where T is determined from the relation $\int_{0}^{T} \varphi(s)ds = \int_{r_0}^{\infty} du/f(u)$.*

If $\int_{0}^{\infty} \varphi(s)ds \le \int_{r_0}^{\infty} du/f(u)$, then $V(t, x(t))$ is defined for all $t : 0 \le t < \infty$ and if $\int_{0}^{\infty} \varphi(s)ds < \int_{r_0}^{\infty} du/f(u)$, then $V(t, x(t))$ is bounded on the interval $[0, \infty)$.

Theorem 2.4. *If $V' \le \varphi(t)f(V)$ and $\int_{r_0}^{\infty} du/g(u) = \infty$, then the function $V(t, x(t))$ is defined for all $0 \le t < \infty$.*

If, moreover, $\int_{t_0}^{t} \varphi(s)ds < \infty$, then the function $V(t, x(t))$ on the solutions of the system (2.1) is limited to the interval $[t_0, \infty)$.

To date, the comparison principle has been developed for many classes of equations in finite-dimensional and infinite-dimensional spaces. The results obtained are summarised in many papers and monographs (see, for example, [4], [66]–[68], [142], and the bibliography therein). At the same time, the lack of a general method for analysing the dynamic properties of solutions to equations and/or comparison systems stimulates the obtaining of new estimates for the variation of Lyapunov functions for certain classes of systems of equations. The estimates given in Theorems 2.1–2.4 are an example of such a search and have some potential for applications.

2.3   Systems of Equations with Quadratic Nonlinearity

Let us assume that the system of nonlinear equations of perturbed motion (2.1) has a quadratic right-hand side. Suppose that in the neighbourhood G of the state $x = 0$ of the system (2.1), there exists a function V that is single-valued, continuous, and definitely positive in the sense of Lyapunov.

Next, we study the system of perturbed motion equations (2.1) with a specific type of nonlinearity, namely a *quadratic nonlinearity* (see [108] and the bibliography therein):

$$\frac{dx}{dt} = A(t)x(t) + X^T(t)B(t)x(t), \quad x(t_0) = x_0. \tag{2.22}$$

Here, $A(t)$ is an $n \times n$ matrix with elements continuous on any finite interval, $B(t)$ is a rectangular $n^2 \times n$ matrix consisting of symmetric square matrices $B_i(t)$, $i = 1, 2, \ldots, n$,

$$B_i(t) = \begin{bmatrix} b^i_{11}(t) & b^i_{12}(t) & \cdots & b^i_{1n}(t) \\ b^i_{21}(t) & b^i_{22}(t) & \cdots & b^i_{2n}(t) \\ \vdots & \vdots & \ddots & \vdots \\ b^i_{n1}(t) & b^i_{n2}(t) & \cdots & b^i_{nn}(t) \end{bmatrix}, \quad i = 1, 2, \ldots, n,$$

and $X^T(t) = \{X_1(t), X_2(t), \ldots, X_n(t)\}$ is a rectangular $n \times n^2$ matrix consisting of square $n \times n$ matrices $X_i(t)$, which have vectors $x(t)$ on the ith rows, while its other elements are equal to zero, i.e.,

$$X_1(t) = \begin{bmatrix} x_1(t) & x_2(t) & \vdots & x_n(t) \\ 0 & 0 & \vdots & 0 \\ \vdots & \vdots & \ddots & \vdots \\ 0 & 0 & \vdots & 0 \end{bmatrix},$$

$$X_2(t) = \begin{bmatrix} 0 & 0 & \vdots & 0 \\ x_1(t) & x_2(t) & \vdots & x_n(t) \\ \vdots & \vdots & \ddots & \vdots \\ 0 & 0 & \vdots & 0 \end{bmatrix},$$

$$\ldots, X_n(t) = \begin{bmatrix} 0 & 0 & \vdots & 0 \\ 0 & 0 & \vdots & 0 \\ \vdots & \vdots & \ddots & \vdots \\ x_1(t) & x_2(t) & \vdots & x_n(t) \end{bmatrix}.$$

The total derivative of the function V corresponding to the system (2.22) is calculated using the formula

$$V'(t,x) = V_t(t,x) + (A(t)x(t) + X^T(t)B(t)x(t), \nabla_x V(t,x)). \qquad (2.23)$$

The purpose of this section is to obtain a new estimate for the Lyapunov function V along solutions of the system (2.22) and the conditions for boundedness of motion, practical boundedness, and convergence of solutions.

2.3.1   Estimation of Lyapunov functions V

The Lyapunov relation (see [76], p. 62) is fundamental in obtaining the integral inequality for a given Lyapunov function mapping $\mathbb{R}_+ \times \mathbb{R}^n \to \mathbb{R}_+$: it connects the function $V(t,x)$ and its derivative $V'(t,x)$ by virtue of the equations of motion (2.22). Let us show that the following statement is true.

Lemma 2.1. *Let the total derivative $V'(t,x)$ have positive functions $f_1(t)$ and $f_2(t)$(integrable over any finite interval) such that*

$$V'(t,x) \le f_1(t)V(t,x) + f_2(t)V^2(t,x) \qquad (2.24)$$

in the range $(t,x) \in \mathbb{R}_+ \times D$, where $D \subseteq \mathbb{R}^n$.

Then, along the solutions of the system (2.22), the estimate

$$V(t,x(t)) \le V(t_0,x_0) \exp\left[\int_{t_0}^t f_1(s)ds\right] (L(t_0,t))^{-1} \qquad (2.25)$$

is satisfied for all $t \ge t_0$, for which

$$L(t_0,t) = 1 - V(t_0,x_0)\int_{t_0}^t f_2(s)\exp\left[\int_{t_0}^s f_1(\tau)d\tau\right]ds > 0.$$

Proof.   In view of the *Lyapunov relation*

$$V(t,x) = V(t_0,x_0) + \int_{t_0}^t V'(s,x(s))ds,$$

the inequality (2.24) implies that

$$V(t,x(t)) \le V(t_0,x_0) + \int_{t_0}^t (f_1(s)V(s,x(s)) + f_2(s)V^2(s,x(s)))ds,$$

$$t \ge t_0. \qquad (2.26)$$

Let $m(t) = V(t,x(t))$ for all $t \ge t_0$ and $m(t_0) = V_0(t_0,x_0)$. From (2.26), we get the estimate

$$m(t) \le m(t_0) + \int_{t_0}^t (f_1(s) + f_2(s)m(s))m(s)ds, \quad t \ge t_0. \qquad (2.27)$$

Applying Lemma 1.4 to the estimate (2.27), we arrive at the inequality

$$m(t) \leq m(t_0) \exp\left[\int_{t_0}^{t} f_1(s)ds\right] (L(t_0, t))^{-1}, \qquad (2.28)$$

where $L(t_0, t) = 1 - m(t_0) \int_{t_0}^{t} f_2(s) \exp\left[\int_{t_0}^{t} f_1(\tau)d\tau\right] ds$.

The inequality (2.28) holds for all $t \geq t_0$, for which $L(t_0, t) > 0$. Given that $m(t) = V(t, x(t))$, the estimate of the function V along the solutions of the system (2.23) is obtained in the form (2.25).

Lemma 2.1 is proved. $\qquad\qquad\qquad\qquad\qquad\qquad\qquad\qquad\qquad$ □

Corollary 2.1. *Let* (2.24) $f_2(t) \equiv 0$ *for all* $t \geq t_0$. *Then,* (2.25) *implies that*

$$V(t, x(t)) \leq V(t_0, x_0) \exp\left[\int_{t_0}^{t} f_1(s)ds\right], \quad t \geq t_0. \qquad (2.29)$$

Corollary 2.2. *Let the function* $f_1(t) \equiv 0$ *in the inequality* (2.24) *for all* $t \geq t_0$. *Then,* (2.25) *implies that* $V(t, x(t)) \leq V(t_0, x_0)(L^*(t_0, t))^{-1}$ *for all* $t \geq t_0$, *for which*

$$L^*(t_0, t) = 1 - V(t_0, x_0) \int_{t_0}^{t} f_2(s)ds > 0.$$

The resulting estimate (2.25) and Corollaries 2.1 and 2.2 can be used in many problems of the general theory of motion stability. Let us consider some of them.

2.3.2 Boundedness theorems

Here and in the following, the vector and matrix norms are taken in the form

$$\|x(t)\| = \left\{\sum_{i=1}^{n} x_i^2(t)\right\}^{1/2}, \quad \|B\| = \left\{\lambda_{\max}(B^T B)\right\}^{1/2},$$

where $\lambda_{\max}(\diamond)$ and $\lambda_{\min}(\diamond)$ are the extreme eigenvalues of the corresponding symmetric matrices. The system (2.22) is considered in the region $\mathbb{R}_+ \times \mathbb{R}^n$, and we assume that the right-hand side of the system is defined and continuous. Following [169], we recall the following definitions.

Definition 2.3. *A solution* $x(t, t_0, x_0)$ *of the system* (2.1) *is* bounded *if there exists a positive constant* $\beta > 0$ *such that* $\|x(t, t_0, x_0)\| < \beta$ *for all* $t \geq t_0$, *where* β *may depend on each solution.*

Definition 2.4. A *solution* to the system (2.22) is *equi-bounded* if for any $\alpha > 0$ and $t_0 \in \mathbb{R}_+$, there exists $\beta(t_0, \alpha) > 0$ such that if $x_0 \in S_\alpha = \{x \in \mathbb{R}^n : \|x_0\| < \alpha\}$, then $\|x(t, t_0, x_0)\| < \beta(t_0, \alpha)$ for all $t \geq t_0$.

Definition 2.5. (see [48]) The function $a : \mathbb{R}_+ \to \mathbb{R}_+$ is continuous, strictly increasing, satisfies $a(0) = 0$, and belongs to the K-class.

Denote by Ω the region in the space $\mathbb{R}^n$ containing the state $x = 0$ of the system (3.1).

The following assertion holds.

Definition 2.6. If $a : \mathbb{R}_+ \to \mathbb{R}_+$, if $a \in K$-class, and if $\lim\limits_{r \to \infty} a(r) = \infty$, then a is said to belong to KR-class.

Theorem 2.5. *Let the system* (2.22) *have a Lyapunov function* $V(t, x)$ *defined on* $\mathbb{R}_+ \times \mathbb{R}^n$ *for which the conditions of Lemma 2.1 are satisfied and:*

(1) $a(\|x\|) \leq V(t, x)$, *where* $a \in KR$-*class;*
(2) $V'(t, x)$ *satisfies the estimate* (2.24) *for all* $(t, x) \in \mathbb{R}_+ \times \mathbb{R}^n$;
(3) $V(t_0, x_0) \exp\left[\int_{t_0}^{t} f_1(s)ds\right] (L(t_0, t))^{-1} < a(\beta)$ *for all* $t \geq t_0$ *for some value* $\beta > 0$.

Then, the solutions $x(t, t_0, x_0)$ *of the* (2.22) *system are equi-bounded.*

Proof. Let $x(t, t_0, x_0)$ be a solution to the system (2.22) such that $x_0 \in S_\alpha$. Under condition (1) of Theorem 2.5, we have $\|x(t)\| \leq a^{-1}(V(t, x(t)))$ for all $t \geq t_0$. A consequence of condition (2) of Theorem 2.5 is the estimate (2.25), which holds for all $t \geq t_0$ for which $L(t_0, t) > 0$. □

Condition (3) implies that $\|x(t)\| \leq a^{-1}(V(t, x(t))) < a^{-1}a(\beta) = \beta(t_0, \alpha)$ for all $t \geq t_0$. This proves the equi-boundedness of the solution $x(t) = x(t, t_0, x_0)$.

Theorem 2.6. *Let there exist a function* $V(t, x)$ *defined on* $\mathbb{R}_+ \times \mathbb{R}^n$, $\|x\| \geq \mathbb{R}$, *where* $\mathbb{R}$ *is an arbitrarily large number that satisfies the following conditions:*

(1) $a(\|x\|) \leq V(t, x) \leq b(\|x\|)$, *where* $a \in KR$-*class*, $b(r) \in K$-*class is a continuous function on* $\mathbb{R}_+$;
(2) *the conditions of Lemma 2.1 hold for all* $(t, x) \in \mathbb{R}_+ \times \mathbb{R}^n$;
(3) *for* $x_0 \in S_\alpha$, $V(t_0, x_0) \leq b(\alpha)$ *and* $\exp\left[\int_{t_0}^{t} f_1(s)ds\right] (L(t_0, t))^{-1} < \frac{a(\beta)}{b(\alpha)}$ *for all* $t \geq t_0$.

Then, the solutions of the system (2.23) *are uniformly bounded.*

Proof. Consider the solution $x(t) = x(t, t_0, x_0)$ of the system (2.22) with the initial conditions $x_0 \in S_\alpha$. Under conditions (1) and (2) of Theorem 2.6, we have the following estimates:

$$\|x(t)\| \leq a^{-1}(V(t, x)) \quad \text{for all} \ t \geq t_0 \quad \text{and}$$

$$V(t, x(t)) \leq V(t_0, x_0) \exp\left[\int_{t_0}^{t} f_1(s)ds\right] (L(t_0, t))^{-1},$$

where $L(t_0, t) > 0$ for all $t \geq t_0$. From the fact that $\exp\left[\int_{t_0}^{t} f_1(s)ds\right] \times (L(t_0, t))^{-1} < \frac{a(\beta)}{b(\alpha)}$, $t \geq t_0$, it follows that the estimate $\|x(t, t_0, x_0)\| < a^{-1}(a(\beta)) = \beta(\alpha)$ holds, where $\beta(\alpha)$ does not depend on t_0. This proves Theorem 2.6.   □

2.3.3   Boundedness of motion of two coupled systems

Consider a system of coupled conditions

$$\frac{dx}{dt} = F(t, x, y), \quad \frac{dy}{dt} = G(t, x, y), \tag{2.30}$$

where $x \in \mathbb{R}^n$, $y \in \mathbb{R}^m$, and the vector functions $F(t, x, y)$ and $G(t, x, y)$ are defined and continuous on $\mathbb{R}_+ \times \mathbb{R}^n \times \mathbb{R}^m$. Suppose that for the system (2.30), the Lyapunov function $V(t, x, y)$ is constructed, for which we establish an estimate similar to the estimate (2.25).

Lemma 2.2. *Let a Lyapunov function* $V(t, x, y)$ *be constructed for the system* (2.30) *such that*

$$V'(t, x, y) \leq \psi_1(t)V(t, x, y) + \psi_2(t)V^2(t, x, y) \tag{2.31}$$

for all $t \in \mathbb{R}_+$, *where* $\psi_1(t)$ *and* $\psi_2(t)$ *are positive integrable functions on any finite interval.*

    Then, the function $V(t, x, y)$ *satisfies the estimate*

$$V(t, x(t), y(t)) \leq V(t_0, x_0, y_0) \exp\left[\int_{t_0}^{t} \psi_1(s)ds\right] (M(t_0, t))^{-1} \tag{2.32}$$

for all $t \geq t_0$, *for which*

$$M(t_0, t) = 1 - V(t_0, x_0, y_0) \int_{t_0}^{t} \psi_2(s) \exp\left[\int_{t_0}^{t} \psi_1(\tau)d\tau\right] ds > 0.$$

The proof of Lemma 2.2 is similar to that of Lemma 2.1. Next, we present some corollaries of Lemma 2.2.

Corollary 2.3. *Let in* (2.31) $\psi_2(t) \equiv 0$ *for all* $t \geq t_0$. *Then,*

$$V(t, x(t), y(t)) \leq V(t_0, x_0, y_0) \exp \int_{t_0}^{t} \psi_1(s) ds \qquad (2.33)$$

for all $t \geq t_0$.

Corollary 2.4. *Let in* (2.31) $\psi_1(t) \equiv 0$ *for all* $t \geq t_0$. *Then,*

$$V(t, x(t), y(t)) \leq V(t_0, x_0, y_0)(M^*(t_0, t))^{-1} \qquad (2.34)$$

for all $t \geq t_0$, *for which*

$$M^*(t_0, t) = 1 - V(t_0, x_0, y_0) \int_{t_0}^{t} \psi_2(s) ds > 0.$$

Theorem 2.7. *Let the system* (2.30) *have a Lyapunov function* $V(t, x, y)$ *defined in the domain* $S(r) = \{(x, y)$ *in* $\mathbb{R}^n \times \mathbb{R}^m : \|x\|^2 + \|y\|^2 \geq r^2\}$ *and whose total derivative satisfies the estimate* (2.31), *where* r *can be an arbitrarily large number. Besides:*

(1) *there exists a function* $a \in KR\text{-class}$ *such that* $a(\|y\|) \leq V(t, x, y)$ *uniformly in* (t, x);
(2) *there exists a continuous function* $b(r, s)$ *such that* $V(t, x, y) \leq b(\|x\|, \|y\|)$;
(3) *for the values* (t_0, x_0, y_0) *such that* $\|x_0\|^2 + \|y_0\|^2 \leq \alpha^2$, $\alpha > r$, *there exists* $\beta(\alpha) > 0$ *such that* $V(t_0, x_0, y_0) \leq b(\alpha)$ *and*

$$\exp\left[\int_{t_0}^{t} \psi_1(s) ds\right] (M(t_0, t))^{-1} < \frac{a(\beta)}{b(\alpha)} \quad \forall \geq t_0.$$

Then, the solutions of the system (2.33) *are uniformly y-bounded.*

Proof. Consider a solution, $(x^T(t, x_0, y_0)$ and $y^T(t, x_0, y_0)^T)$, of the system (2.30) with the initial conditions (t_0, x_0, y_0) such that $\|x_0\|^2 + \|y_0\|^2 \leq \alpha^2$, $\alpha > r$. For a function $V(t, x, y)$ satisfying the condition (1) of Theorem 2.7 along the solutions of the system (2.30), we have the estimate

$$\|y(t)\| \leq a^{-1}(V(t, x, y)) \quad \forall \geq t_0. \qquad (2.35)$$

$\square$

Further, from condition (2), for the function $V(t, x, y)$, the inequality

$$V(t, x(t), y(t)) \leq b(\|x_0\|, \|y_0\|) \exp\left[\int_{t_0}^{t} \psi_1(s)ds\right] (M(t_0, t))^{-1} \qquad (2.36)$$

holds for all $t \geq t_0$. It follows from the estimate (2.36) of the condition (3) and the inequality (2.35) that for all $t \geq t_0$, the estimate $\|y(t)\| < a^{-1}(a(\beta)) = \beta(\alpha)$ is valid. This proves Theorem 2.7.

Theorem 2.8. *Assume that for the system* (2.30), *there exists a Lyapunov function* $V(t, x, y)$ *defined on the set* $t_0 \leq t < \infty$, $\|x\| < \infty$, *and* $\|y\| \geq k > 0$ *and satisfying the following conditions:*

(1) $a(\|y\|) \leq V(t, x, y) \leq b(\|y\|)$, *where* $a, b \in KR$-*class;*
(2) $V'(t, x, y) \leq \psi_1(t)V(t, x, y) + \psi_2(t)V^2(t, x, y)$, *where* $\psi_1(t), \psi_2(t)$ *are positive integrable functions on any finite interval;*
(3) $\exp\left[\int_{t_0}^{t} \psi_1(s)ds\right] (M(t_0, t))^{-1} < \frac{a(\beta)}{b(\alpha)}$ *for all* $t \geq t_0$.

Moreover, for every $m > 0$, *there is a Lyapunov function* $W(t, x, y)$ *defined on the set* $t_0 \leq t < \infty$, $\|x\| \geq K_1(m)$, *and* $\|y\| \geq m$ *and satisfying the following conditions:*

(4) $a_1(\|x\|) \leq W(t, x, y) \leq b_1(\|x\|)$, *where* $a_1, b_1 \in KR$-*class;*
(5) $W'(t, x, y) \leq \varphi_1(t)W(t, x, y) + \varphi_2(t)W^2(t, x, y)$, *where* $\varphi_1(t), \varphi_2(t)$ *are positive integrable functions on any finite interval;*
(6) $\exp\left[\int_{t_0}^{t} \varphi_1(s)ds\right] (M^*(t_0, t))^{-1} < \frac{a_1(\beta_1)}{b_1(\alpha_1)}$ *for all* $t \geq t_0$, *where*

$$M^*(t_0, t) = 1 - W(t_0, x_0, y_0)\int_{t_0}^{t} \varphi_2(s)\exp\left[\int_{t_0}^{t}\varphi_1(\tau)d\tau\right]ds > 0.$$

Then, the solutions of the system (2.30) *are uniformly bounded.*

Proof. Let $\alpha > 0$ be given such that $K < \alpha$. Consider the solution $(x(t), y(t))^T$ of the system (2.30) under the initial conditions $t_0 \geq 0$, $\|x_0\| \leq \alpha$ and $\|y_0\| \leq \alpha$. We choose $\beta(\alpha)$ so large that $b(\alpha) < a(\beta)$ and show that under conditions (1)–(3) of Theorem 2.8, $\|y(t)\| < \beta(\alpha)$ for all $t \geq t_0$, for which the solution $(x(t), y(t))^T$ of the system (2.30) exists. Under condition (2) of Theorem 2.8, according to Lemma 2.2, we have the estimate (2.32), from which it follows that

$$V(t, x(t), y(t)) \leq b(\alpha)\exp\left[\int_{t_0}^{t}\psi_1(s)ds\right](M(t_0, t))^{-1}, \quad \forall \geq t_0, \qquad (2.37)$$

since, by condition (1), $V(t_0, x_0, y_0) \leq b(\alpha)$. From the estimate (2.37) and condition (3) of Theorem 2.8, we obtain

$$\|y(t)\| \leq a^{-1}(V(t, x, y)) < a^{-1}(a(\beta)) = \beta(\alpha) \qquad (2.38)$$

for all $t \geq t_0$, for which solutions of the system (2.30) exist. $\qquad \square$

Let $\alpha_1(\alpha) = \max\{\alpha, K_1(\beta(\alpha))\}$, and consider the function $W(t, x, y)$ in the domain $t_0 \leq t < \infty$, $\|x\| \geq K_1(\beta)$, $\|y\| \leq \beta$. Let us choose $\beta_1(\alpha)$ so large that $b_1(\alpha_1) < a_1(\beta_1)$ and show that under conditions (4)–(6) of Theorem 2.8, we have the estimate $\|x(t)\| < \beta_1(\alpha)$ for all $t \geq t_0$, for which a solution to the system (2.30) exists. From the condition (5) of Theorem 2.8, according to Lemma 2.2, it follows that

$$W(t, x(t), y(t)) \leq W(t_0, x_0, y_0) \exp\left[\int_{t_0}^{t} \varphi_1(s)ds\right] (M^*(t_0, t))^{-1} \qquad (2.39)$$

for all $t \geq t_0$, for which

$$M^*(t_0, t) = 1 - W(t_0, x_0, y_0) \int_{t_0}^{t} \varphi_2(s) \exp\left[\int_{t_0}^{t} \varphi_1(\tau)d\tau\right] ds > 0.$$

From the condition (2.39), it follows that $W(t_0, x_0, y_0) < b_1(\alpha_1)$; therefore, the condition (6) leads to the estimate

$$\|x(t)\| \leq a_1^{-1}(W(t, x(t), y(t))) < a_1^{-1}(a_1(\beta_1)) = \beta_1(\alpha). \qquad (2.40)$$

It follows from the estimation of the solution components of the system (2.30) that $\|x(t)\| < \beta_1(\alpha)$ and $\|y(t)\| < \beta(\alpha)$ for all $t \geq t_0$, for which the solutions exist and the conditions $M(t_0, t) > 0$ and $M^*(t_0, t) > 0$ are satisfied. This proves the theorem.

2.3.4 Practical boundedness of motion

Let us continue with the study of the properties of motion of the system (2.22).

In the space R^n, we define open subsets $S_0(t)$ and $S(t)$ for all $t \geq t_0$. Denote the boundary of $S(t)$ by $\partial S(t)$ and its closure by $\bar{S}(t)$. Further, it is assumed that $\partial S_0(t) \cap \partial S(t) = \varnothing$ for all $t \geq t_0$.

Definition 2.7. The motion of the system (2.22) is *practically bounded* [14] with respect to $\{S_0(t), S(t), t_0\}$ if the solution $x(t, t_0, x_0)$ with the initial condition $x_0 \in S_0(t_0)$ satisfies $x(t, t_0, x_0) \in S(t)$ for all $t \geq t_0$.

Definition 2.8. The motion of the system (2.22) is almost *uniformly bounded* with respect to $\{S_0(t), S(t)\}$ if for any $t_i \geq t_0$, $x_i \in S_0(t_i)$, the solution $x(t, t, x_i) \in S(t)$ for all $t \in [t_i, \infty)$.

The following assertion holds.

Theorem 2.9. *Assume that for the system* (2.22), *there exists a continuous real function $V(t,x)$ that is locally Lipschitz in x and the functions $\nu_1(t)$ and $\nu_2(t)$ are positive integrable on any finite interval so that:*

(1) $|V(t,x') - V(t,x'')| \leq L\|x' - x''\|$, *where $L > 0$ for all $(x',x'') \in S(t)$ and $t \geq t_0$;*

(2) $V'(t,x) \leq \nu_1(t)V(t,x) + \nu_2(t)V^2(t,x)$ *for all $t \geq t_0$ and $x \in S(t)$;*

(3) $\exp\left[\int_{t_0}^{t} \nu_1(s)ds\right] (K(t_0,t))^{-1} < \dfrac{\inf\limits_{x \in S(t)} V(t,x)}{\sup\limits_{x_0 \in S_0(t_0)} V(t_0,x_0)}$

and $K(t_0,t) = 1 - V(t_0,x_0)\int_{t_0}^{t} \nu_2(s)\exp\left[\int_{t_0}^{s} \nu_1(\tau)d\tau\right] ds > 0$ *for all $t \geq t_0$.*

Then, the motion of the system (2.22) *is practically bounded. If condition* (3) *is replaced by the condition*

(3') $\exp\left[\int_{t_1}^{t_2} \nu_1(s)ds\right] (K(t_1,t_2))^{-1} < \dfrac{\inf\limits_{x \in \partial S(t_2)} V(t_2,x)}{\sup\limits_{x_1 \in \partial S_0(t_1)} V(t_1,x_1)}$

and $K(t_1,t_2) = 1 - V(t_1,x_1)\int_{t_1}^{t_2} \nu_2(s)\exp\left[\int_{t_1}^{s} \nu_1(\tau)d\tau\right] ds > 0$ *for any $t_2 > t_1$, $t_1, t_2 \in [t_0, \infty)$, then the motion of the system* (2.22) *is uniformly practically bounded.*

Proof. Let $x(t,t_0,x_0)$ be the solution of the system (2.22) with the initial conditions $x_0 \in S_0(t_0)$. Suppose that under conditions (1)–(3) of Theorem 2.9, there exists $t_2 \in (t_0, \infty)$ such that $x(t_2,t_0,x_0)\bar{\in}S(t_2)$. From the fact that $S_0(t) \subset S(t)\forall t \geq t_0$ and $\partial S_0(t) \cap \partial S(t) = \varnothing \forall \geq t_0$, it follows that there exists $t_1 > t_0$ such that $x(t,t_0,x_0) \in S(t)$ for all $t \in [t_0,t_1)$ and $x(t_1,t_0,x_0) \in \partial S(t_1)$. Under conditions (1) and (2) of Theorem 2.9, according to Lemma 2.2, we have the estimate

$$V(t,x(t)) \leq V(t_0,x)\exp\left[\int_{t_0}^{t} \nu_1(s)ds\right] (K(t_0,t))^{-1}$$

for all $t \geq t_0$. Since $\sup\limits_{x \in S_0(t_0)} V(t_0,x) \geq V(t_0,x_0)$, by virtue of condition (3) of Theorem 2.9, we obtain

$$V(t_1,x(t),t_0,x_0) < \sup\limits_{x \in S_0(t_0)} V(t_0,x)\exp\left[\int_{t_0}^{t} \nu_1(s)ds\right] (K(t_0,t))^{-1}$$

$$< \inf\limits_{x \in \partial S(t_1)} V(t_1,x).$$

It follows from this inequality that $x(t_1, t_0, x_0) \bar{\in} \partial S(t_1)$, which contradicts the above assumption. Therefore, $x(t, t_0, x_0) \in S(t)$ for all $t \geq t_0$ as soon as $x_0 \in S_0(t_0)$.

If condition (3') is satisfied, arguments similar to those given above lead to the assertion that $x(t, t_1, x_0) \bar{\in} S(t_1)$ for any $t_1 \in (t_0, \infty)$. This proves Theorem 2.9. □

Example 2.1. Let the vector function $f(t, x)$ in the system (2.1) have the form

$$f(t, x) = P(t)x + g(t, x),$$

where $P(t)$ is an $n \times n$ matrix with elements continuous on any finite interval and $g : \mathbb{R}_+ \times \mathbb{R}^n \to \mathbb{R}^n$. Consider the system of equations

$$\frac{dx}{dt} = P(t)x + g(t, x), \tag{2.41}$$

$$x(t_0) = x_0. \tag{2.42}$$

Let $\lambda_M(t)$ and $\lambda_m(t)$ denote, respectively, the maximum and minimum eigenvalues of the matrix

$$C(t) = \frac{1}{2}(P^T(t) + P(t)),$$

and we apply the Lyapunov function $V(t, x) = V(x) = x^T x$. It is easy to show that

$$V'(x) \leq \lambda_M(t)V(x) + x^T g(t, x). \tag{2.43}$$

Let there exist a number $\mathbb{R} > 0$ such that the scalar function $k(t)$, continuous for all $t \geq t_0$, satisfies the estimate

$$x^T g(t, x) \leq k(t)(x^T x)^2 \tag{2.44}$$

for all $t \geq t_0$ and $\|x\| \geq \mathbb{R}$. Then, from the inequality (2.43), under the condition (2.44), we obtain the inequality

$$V'(x) \leq \lambda_M(t)V(x) + k(t)V^2(x),$$

from which, according to Lemma 2.2, it follows that

$$V(x(t)) \leq V(x_0) \exp\left[\int_{t_0}^t \lambda_M(s)ds\right](N(t_0, t))^{-1} \tag{2.45}$$

for all $t \geq t_0$, for which

$$N(t_0, t) = 1 - V(x_0)\int_{t_0}^t k(s)\exp\left[\int_{t_0}^s \lambda_M(\tau)d\tau\right]ds > 0.$$

Applying Theorem 2.9 to the inequality (2.45), it is easy to obtain boundedness conditions for the solutions of the system (2.41) under the initial conditions x_0 such that $x_0 \in S_\alpha$.

2.3.5 Convergence of solutions

Consider the system (2.22) with constant matrices $A(t)$ and $B(t)$:

$$\frac{dx}{dt} = Ax(t) + X^T(t)Bx(t). \tag{2.46}$$

Let the matrix A of the linear part of the system (2.46) be asymptotically stable. Then, following the theory of stability in linear approximation, the zero solution of the nonlinear system (2.46) will also be asymptotically stable. As the Lyapunov function, we choose the quadratic form $V(x) = x^T H x$, where H is a constant $n \times n$ matrix, and calculate its total derivative by virtue of the system (2.46):

$$V'(x(t)) = [Ax(t) + X^T(t)Bx(t)]^T Hx(t) + x^T(t)H[Ax(t) + X^T(t)Bx(t)]$$
$$= x^T(t)[(A^T H + HA) + (B^T X(t)H + HX^T(t)B)]x(t). \tag{2.47}$$

Since by assumption the matrix A is asymptotically stable, for an arbitrary positive definite matrix C, the *Lyapunov matrix equation*

$$A^T H + HA = -C$$

has a unique solution: a positive definite matrix H. Taking into account that H is a solution of the Lyapunov equation from (2.47), we obtain

$$V'(x(t)) = -x^T(t)[C - (B^T X(t)H + HX^T(t)B)]x(t). \tag{2.48}$$

The stability region of the zero equilibrium of the system (2.46) is the interior of the level surface of the Lyapunov function $V(x) = r > 0$, which lies inside the region

$$G_0 = \{x \in \mathbb{R}^n : C - B^T XH - HX^T B > \Theta\},$$

where, under the symbol

$$C - B^T XH - HX^T B > \Theta, \tag{2.49}$$

the positive definiteness of the matrix H is understood. Let us change the condition (2.48) to be more rough. Since due to the chosen matrix and vector norms, the relation

$$\|X(t)\| = \|x(t)\|$$

holds, the total derivative of the Lyapunov function $V(x) = x^T H x$ will satisfy the estimate

$$V'(x(t)) \leq -[\lambda_{\min}(C) - 2\|H\|\|B\|\|x(t)\|^2].\tag{2.50}$$

Denote

$$G_0 = \left\{ x \in \mathbb{R}^n : \|x\| < \frac{\lambda_{\min}(C)}{2\|H\|\|B\|} \right\}.\tag{2.51}$$

Then, the region of "guaranteed" stability takes the form

$$G_{r_0} = \max_{r>0} \{G_r : G_r \subset G_0\}, \quad G_r = \left\{ x \in \mathbb{R}^n : x^T H x < r^2 \right\}.\tag{2.52}$$

Following this dependence, to determine the "maximum" region of stability, one should "embed" the ellipsoid $x^T H x = r^2$ inside a sphere of radius $\mathbb{R} = \frac{\lambda_{\min}(C)}{2\|H\|\|B\|}$ and "stretch" it at $r \to \infty$ until the ellipse touches the sphere.

Theorem 2.10. *Let the matrix of the linear part of the system* (2.46) *be asymptotically stable. Then, the zero solution of the system* (2.46) *is also asymptotically stable; for its solutions satisfying the initial conditions*

$$\|x(0)\| < \frac{\gamma(H)}{2\|B\|\varphi(H)}, \quad \varphi(H) = \frac{\lambda_{\max}(H)}{\lambda_{\min}(H)}, \quad \gamma(H) = \frac{\lambda_{\min}(C)}{\lambda_{\max}(H)},\tag{2.53}$$

the following convergence estimate is valid:

$$\|x(t)\| \leq \frac{\gamma(H)\sqrt{\lambda_{\min}(H)}|x(0)|}{[\gamma(H) - 2\|B\|\varphi(H)\|x(0)\|]e^{\frac{1}{2}\gamma(H)t} + 2\|B\|\varphi(H)\|x(0)\|}.\tag{2.54}$$

Proof.   An estimate for the *convergence of solutions* in the stability region can be obtained using the quadratic Lyapunov function $V(x) = x^T H x$. Its total derivative due to the system (2.46) is estimated using the inequality (2.50). Since the quadratic function $V(x) = x^T H x$ satisfies the two-sided inequality

$$\lambda_{\min}(H)\|x\|^2 \leq V(x) \leq \lambda_{\max}(H)\|x\|^2,\tag{2.55}$$

the inequality (2.50) can be rewritten as

$$V'(x(t)) \leq -\frac{\lambda_{\min}(C)}{\lambda_{\max}(H)}V(x(t)) + 2\lambda_{\max}(H)\|B\|\frac{V^{3/2}x(t)}{\lambda_{\min}^{3/2}(H)}.\tag{2.56}$$

$\square$

Using the notation (2.53), we rewrite the resulting expression in the form

$$V'(x(t)) \leq -\gamma(H)V(x(t)) + 2\frac{\|B\|\varphi(H)}{\sqrt{\lambda_{\min}(H)}}V^{3/2}(x(t)).$$

We divide it by $V^{3/2}(x)$ to get the estimate

$$V^{3/2}(x)V'(x(t)) \leq -\gamma(H)V^{-1/2}(x(t)) + 2\frac{\|B\|\varphi(H)}{\sqrt{\lambda_{\min}(H)}}.$$

Denoting $V^{-1/2}x(t) = z(t)$, we get

$$-2\frac{dz(t)}{dt} \leq -\gamma(H)z(t) + 2\frac{\|B\|\varphi(H)}{\sqrt{\lambda_{\min}(H)}}.$$

From here,

$$\frac{dz(t)}{dt} \geq \frac{1}{2}\gamma(H)z(t) - \frac{\|B\|\varphi(H)}{\sqrt{\lambda_{\min}(H)}}.$$

Solving the resulting inequality (by analogy with the linear *inhomogeneous Bernoulli equation*), we obtain

$$z(t) \geq \left[z(0) - 2\frac{\|B\|\varphi(H)}{\gamma(H)\sqrt{\lambda_{\min}(H)}}\right]e^{\frac{1}{2}\gamma(H)t} + 2\frac{\|B\|\varphi(H)}{\gamma(H)\sqrt{\lambda_{\min}(H)}},$$

and taking into account that $V^{-1/2}x(t) = z(t)$, we find

$$V^{-1/2}x(t)$$
$$\geq \left[V^{-1/2}x(0) - 2\frac{\|B\|\varphi(H)}{\gamma(H)\sqrt{\lambda_{\min}(H)}}\right]e^{\frac{1}{2}\gamma(H)t} + 2\frac{\|B\|\varphi(H)}{\gamma(H)\sqrt{\lambda_{\min}(H)}}.$$

From here,

$$V^{-1/2}x(t)$$
$$\leq \left\{\left[V^{-1/2}x(0) - 2\frac{\|B\|\varphi(H)}{\gamma(H)\sqrt{\lambda_{\min}(H)}}\right]e^{\frac{1}{2}\gamma(H)t} + 2\frac{\|B\|\varphi(H)}{\gamma(H)\sqrt{\lambda_{\min}(H)}}\right\}^{-1}.$$

Using the two-sided inequalities of quadratic forms (2.55), we obtain

$$
\sqrt{\lambda_{\min}(H)}\|x(t)\|
$$

$$
\leq \left\{ \left[\frac{1}{\sqrt{V(x(0))}} - 2\frac{\|B\|\varphi(H)}{\gamma(H)\sqrt{\lambda_{\min}(H)}} \right] e^{\frac{1}{2}\gamma(H)t} \right.
$$

$$
\left. + 2\frac{\|B\|\varphi(H)}{\gamma(H)\sqrt{\lambda_{\min}(H)}} \right\}^{-1}
$$

$$
\leq \left\{ \left[\frac{1}{\sqrt{\lambda_{\min}(H)}\|x(0)\|} - 2\frac{\|B\|\varphi(H)}{\gamma(H)\sqrt{\lambda_{\min}(H)}} \right] e^{\frac{1}{2}\gamma(H)t} \right.
$$

$$
\left. + 2\frac{\|B\|\varphi(H)}{\gamma(H)\sqrt{\lambda_{\min}(H)}} \right\}^{-1}
$$

$$
= \frac{\gamma(H)\sqrt{\lambda_{\min}(H)}\|x(0)\|}{[\gamma(H) - 2\|B\|\varphi(H)\|x(0)\|]e^{\frac{1}{2}\gamma(H)t} + 2\|B\|\varphi(H)\|x(0)\|}.
$$

Thus, for the solutions $x(t)$ of the system (2.46) with the initial conditions in the region (2.53), i.e., $x_0 \in G_0$, the convergence estimate (2.54) is valid.

Example 2.2. Consider the first-order scalar equation

$$
\dot{x}(t) = -ax(t) + bx^2(t), \quad x(0) = x_0, \tag{2.57}
$$

where $a > 0$. This equation is a separable variable equation. Its exact solution is the function

$$
x(t) = \frac{x_0 e^{-at}}{a - x_0[1 - e^{-at}]}. \tag{2.58}
$$

Consider the application of the Lyapunov function method with the function $V(x) = x^2$ for the equation (2.57). For this function, $\lambda_{\max}(H) = \lambda_{\min}(H) = 1$. The total derivative due to the linear part has the form

$$
V'(x(t)) = -2ax^2(t).
$$

Therefore, $\varphi(H) = 1$ and $\gamma(H) = 2a$. The convergence estimate (2.54) for solutions with the initial conditions $|x_0| < a/b$ has a similar form

$$
|x(t)| \leq \frac{a|x(0)|}{[a - |b|\|x(0)\|]e^{at} + |b|\|x(0)\|} = \frac{a|x(0)|e^{-at}}{a - |b||x(0)|[1 - e^{-at}]} \to 0.
$$

Thus, for the scalar equation (2.57) with the exact solution (2.58), the convergence estimate coincides with the estimate obtained by applying the quadratic Lyapunov function.

2.4 General Case of a Quasi-linear System

We consider a *quasi-linear system* of differential equations of perturbed motion

$$\frac{dx}{dt} = P(t)x + g(t, x), \tag{2.59}$$

$$x(t_0) = x_0, \tag{2.60}$$

where $x \in \mathbb{R}^n$, $P(t)$ is an $n \times n$-matrix with elements continuous on any finite interval and $g : \mathbb{R}_+ \times \mathbb{R}^n \to \mathbb{R}^n$, and the solution to the problem (2.59)–(2.60) exists on any finite interval $J = [t_0, t_0 + a)$, $a = \text{const} < \infty$.

Assume that the Lyapunov function $V(t, x)$ is constructed for the system (2.59). One of the questions that arise in the application of the direct Lyapunov method is the question of estimating the variation of the function $V(t, x)$ along the solutions of the system (2.59). Next, we present one way to obtain such an estimate based on the method of integral inequalities.

2.4.1 Lyapunov function estimation

Let the function V be definitely positive and locally Lipschitz. Let us define its total derivative $V'(t, x)$ along the solutions of the system (2.59) for all $(t, x) \in J \times D$, where $D \subseteq \mathbb{R}^n$.

Let us show that the following assertion holds.

Lemma 2.3. *Let the perturbed motion equations (2.59) be such that:*

(1) *there exists the function $V(t, x)$ specified above;*
(2) *there are non-negative integrable functions $a(t)$ and $b(t)$ and a constant $k > 1$ such that*

$$V'(t, x) \le a(t)V(t, x) + b(t)V^k(t, x) \tag{2.61}$$

for all $(t, x) \in J \times D$;
(3) *for all $t \in J^* \subseteq J$,*

$$L(t_0, t) = 1 - (k - 1)V^k(t_0, x_0)$$

$$\times \int_{t_0}^{t} b(s) \exp\left((k - 1) \int_{t_0}^{s} a(\tau)d\tau \right) ds > 0. \tag{2.62}$$

Then, along the solutions of the system (2.59), the estimate

$$V(t, x(t)) \leq V(t_0, x_0) \exp \int_{t_0}^{t} a(s)ds(L(t_0, t))^{-\frac{1}{k-1}} \qquad (2.63)$$

holds for all $t \in J^*$.

Proof. From the inequality (2.64), we get

$$V(t, x(t)) \leq V(t_0, x_0) + \int_{t_0}^{t} \left(a(s)V(s, x(s)) + b(s)V^k(s, x(s)) \right) ds \quad (2.64)$$

for all $t \in J$. We represent the inequality (2.64) as

$$V(t, x(t)) \leq V(t_0, x_0) + \int_{t_0}^{t} \left(a(s) + b(s)V^{k-1}(s, x(s)) \right) V(s, x(s))ds, \quad (2.65)$$

$\square$

from which it follows that

$$V(t, x(t)) \leq V(t_0, x_0) \exp \left[\int_{t_0}^{t} \left(a(s) + b(s)V^{k-1}(s, x(s)) \right) ds \right] \qquad (2.66)$$

for all $t \in J^*$.

To evaluate the multiplier

$$\exp \left[\int_{t_0}^{t} b(s)V^{k-1}(s, x(s))ds \right] \qquad (2.67)$$

on the right-hand side of the inequality (2.63), we apply some results from [72, 83].

As a result, we obtain the estimate

$$V^{k-1}(t, x(t)) \leq V^{k-1}(t_0, x_0) \exp \left[(k-1) \int_{t_0}^{t} a(s)ds \right] (L(t_0, t))^{-1},$$

from which, taking into account that $k > 1$ and $V(t_0, x_0) > 0$, we obtain the estimate (2.63) under the condition (2.62). This proves Lemma 2.3.

Corollary 2.5. *Let, in the estimate* (2.61), *the function* $b(t) = 0$ *for all* $t \in J$. *Then, the estimate* (2.63) *becomes*

$$V(t, x(t)) \leq V(t_0, x_0) \exp \left(\int_{t_0}^{t} a(s)ds \right) \qquad (2.68)$$

for all $t \in J$.

Corollary 2.6. *Let, in the bound* (2.61), *the function* $a(t) = 0$ *for all* $t \in J$. *Then, the estimate* (2.63) *becomes*

$$V(t, x(t)) \leq V(t_0, x_0)(L^*(t_0, t))^{-\frac{1}{k-1}} \qquad (2.69)$$

for all $t \in J^* \subseteq J$, for which

$$L^*(t_0, t) = 1 - (k-1)V^{k-1}(t_0, x_0) \int_{t_0}^{t} b(s)ds > 0.$$

2.4.2 Stability conditions on a finite interval

Consider the problem of *stability on a finite interval* of the system (2.59). Taking into account the results of [81, 176], we present the following definition.

Definition 2.9. Given a function $V(t, x)$, constant $0 < c_1 < c_2$, and an interval $J = [t_0, t_0 + \tau]$, the system (2.59) is *stable on a finite interval* if under the initial conditions (2.60), the condition $V(t_0, x_0) \leq c_1$ implies the estimate

$$V(t, x(t)) < c_2$$

for all $t \in J$ along any solution $x(t) = x(t, t_0, x_0)$ of the system (2.62).

Remark 2.1. The domain $V(t, x) = c$, $c \in (0, H)$, $t \geq t_0$, was introduced in [34]. In [48], a similar area was used to obtain conditions for the asymptotic stability and instability of the zero solution of a system of the form (2.59).

Theorem 2.11. *Let the system (2.59) satisfy the conditions of Corollary 2.5, and for all $t \in J$,*

$$\int_{t_0}^{t} a(s)ds < \ln\left(\frac{c_2}{c_1}\right).$$

Then, the system (2.59) is stable on a finite interval in the sense of Definition 2.9.

Proof. Let the solution $x(t)$ leave the domain $G_0 = \{V(t, x_0) \leq c_1\}$ and, at the moment $t^* \in J$, leave the domain $G = \{V(t^*, x) \leq c_2\}$.

It follows from the inequality (2.68) that at the moment $t = t^*$, the following condition holds:

$$c_1 \exp\left(\int_{t_0}^{t^*} a(s)ds\right) = c_2,$$

whence it follows that

$$\int_{t_0}^{t^*} a(s)ds = \ln\left(\frac{c_2}{c_1}\right).$$

This relation contradicts the condition of Theorem 2.11; consequently, the assumption about the existence of a moment $t^* \in J$ for which $x(t^*)$ leaves the domain G is not true. Theorem 2.11 is proved.   □

Theorem 2.12. *Let the system* (2.59) *satisfy the conditions of Corollary 2.6 and, in addition:*

(1) *for all $t \in J^*$, the following inequalities are satisfied:*

$$L_1^*(t_0, t) = 1 - (k-1)c_1^{k-1} \int_{t_0}^{t} b(s)ds > 0;$$

(2)

$$[L_1^*(t_0, t)]^{-\frac{1}{k-1}} < \frac{c_2}{c_1}.$$

Then, the system (2.59) *is stable on a finite interval in the sense of Definition 2.9.*

Proof of Theorem 2.12 is similar to that of Theorem 2.11.

Theorem 2.13. *Let the system* (2.59) *satisfy all the conditions of Lemma 2.3 and, in addition,*

$$\exp\left(\int_{t_0}^{t} a(s)ds\right)[L(t_0, t)]^{-\frac{1}{k-1}} < \frac{c_2}{c_1}.$$

Then, the system (2.59) *is stable on a finite interval in the sense of Definition 2.9.*

Proof of Theorem 2.13 is similar to that of Theorem 2.11.

Example 2.3. Consider the second-order equation (see [70, p. 139])

$$\ddot{x} + p(t)\dot{x} + [a^2 + q(t)]x = 0, \quad a = \text{const} \neq 0, \tag{2.70}$$

where $p(t) \geq 0$ for all $t \in J$ and $\int_{t_0}^{\infty} q(s)ds < +\infty$; the functions $p(t)$ and $q(t)$ are continuous on J.

We rewrite the equation (2.70) as a system

$$\begin{cases} dx/dt = y, & x(t_0) = x_0, \\ dy/dt = -p(t)y - [a^2 + q(t)]x, & y(t_0) = y_0, \end{cases} \tag{2.71}$$

and for the function $V(x, y) = a^2 x^2 + y^2$, we get

$$V'(x(t), y(t)) = -2p(t)y^2(t) - 2q(t)x(t)y(t) \leq 2|q(t)||x(t)y(t)|$$

$$\leq \frac{|q(t)|}{|a|} \left(a^2 x^2(t) + y^2(t) \right) = \overline{a}(t)V(x(t), y(t)). \quad (2.72)$$

Applying Theorem 2.11 to the inequality (2.72), we find that the system (2.71) is stable on a finite interval if

$$\int_{t_0}^{t} \overline{a}(s)ds < \ln\left(\frac{c_2}{c_1}\right) \quad \text{for all} \quad t \in J. \quad (2.73)$$

Here, $0 < c_1 < c_2$ are predefined values.

Example 2.4. In the system (2.59), the vector function $g(t, x)$ is an n-fold series that converges absolutely in D. Using the function $V = x^T x$, we obtain the estimate

$$V'(x(t)) \leq \lambda_M(t)V(x(t)) + R(t)V(x(t)), \quad (2.74)$$

where $\lambda_M(t)$ is the maximum eigenvalue of the matrix $\frac{1}{2}\left(A^T(t) + A(t)\right)$, and the continuous non-negative function $R(t)$ is defined by the expression (see [81])

$$R(t) = \max_{V(x)=c_2} \frac{|x^T g(t, x)|}{x^T x}.$$

Applying Theorem 2.11 to the inequality (2.74), we obtain the stability conditions on a finite interval of the system (2.59) in the form

$$\int_{t_0}^{t} (\lambda_M(s) + R(s))ds < \ln\left(\frac{c_2}{c_1}\right)$$

for all $t \in J$.

Example 2.5. Consider the system (2.59) under other assumptions about the estimate of the total derivative of the function $V(x) = x^T x$. Since the expression $W(t, x) \equiv x^T g(t, x)$ is a holomorphic function, it is easy to obtain the estimate

$$V'(x(t)) \leq \lambda_M(t)V(x(t)) + \chi(t)V^k(x(t)), \quad (2.75)$$

where $k > 1$ (if the expansion of the function $W(t, x)$ begins with the terms of the third, fourth, etc. order, and $\chi(t)$ is a continuous bounded function).

Applying Theorem 2.13 to the inequality (2.75), we obtain stability conditions on a finite interval in the form of two inequalities:

(1)

$$L_2^*(t_0, t) = 1 - (k-1)c_1^{k-1} \int_{t_0}^{t} \chi(s) \exp\left[(k-1) \int_{t_0}^{s} \lambda_M(\tau)d\tau\right] ds > 0$$

for all $t \in J^*$;

(2)

$$\frac{\exp \int_{t_0}^{t} \lambda_M(s)ds}{[L_2^*(t_0, t)]^{\frac{1}{k-1}}} < \frac{c_2}{c_1}$$

for all $t \in J^*$.

2.5 Systems with Asymptotic Expansion

In this section, nonlinear systems of equations of perturbed motion with an *asymptotic expansion* of the right-hand side are considered. Here, new constructive estimates for the variation of the Lyapunov functions along the solutions of the systems of equations under consideration are obtained. As applications, the problem of the stability of solutions on a finite interval and the problem of estimating approximate integrations are considered.

2.5.1 Preliminary results

Denote by $\mathbb{R}^n$ the real Euclidean space of dimension n, and take as the norm the quantity $\|x\| = \left(\sum_{k=1}^{n} |x_k|^2\right)^{1/2}$ for any $x_k \in \mathbb{R}$.

Let $D_1 \subset \mathbb{R}^n$ be an open n-dimensional domain in the space $\mathbb{R}^n$ and $\mathbb{R} \times \mathbb{R}^n$ be the direct Cartesian product of $\mathbb{R} = (-\infty, \infty)$ and $\mathbb{R}^n$. The symbol (x, y) denotes the scalar product of two vectors $x \in \mathbb{R}^n$ and $y \in \mathbb{R}^n$. For an infinite power series $\sum_{k=0}^{\infty} \mu^k x_k(t, \mu)$, where $\mu \in M = [0, 1)$, μ is a small parameter, and each member of which is defined in the domain $D_2 = \{(t, \mu) : t \in [0, T]\}$, $0 \leq \mu \leq \mu^* < 1$, denote the partial sum by $x_p(t, \mu) = \sum_{k=0}^{p} \mu^k x_k(t, \mu)$.

If in the region $D_1 \times D_2$, there exists a function $x(t, \mu)$ such that

$$\lim\{(x(t, \mu) - x_p(t, \mu))\mu^{-p} : \mu \to 0^+\} = 0,$$

then this power series is an *asymptotic representation* of the function $x(t, \mu)$.

Consider the equations of perturbed motion of a mechanical system of the form

$$\frac{dx}{dt} = f(t, x, \mu),$$ (2.76)

$$x(t_0) = x_0,$$ (2.77)

where $x \in \mathbb{R}^n$ and $f \in C(\mathbb{R}_+ \times \mathbb{R}^n \times M, \mathbb{R}^n)$.

Suppose that the right-hand side of the system (2.76) is defined and continuous in t in the open domain $D \subseteq (\mathbb{R}_+ \times \mathbb{R}^n \times M)$ and has an *asymptotic expansion*:

$$f(t, x, \mu) = \mu f_1(t, x) + \mu^2 f_2(t, x) + \cdots + \mu^m f_m(t, x) = \sum_{k=1}^m \mu^k f_k(t, x).$$ (2.78)

Here, the functions $f_i \in C(\mathbb{R}_+ \times \mathbb{R}^n, \mathbb{R}^n)$, $i = 1, 2, \ldots$, are bounded in t, together with their partial derivatives.

Given the decomposition (2.78), the system (2.76) takes the form

$$\frac{dx}{dt} = \sum_{k=1}^m \mu^k f_k(t, x),$$ (2.79)

with the initial conditions for the solution $x(t)$ in the form (2.77).

Along with the system of equations (2.76), we consider the system of differential equations

$$\frac{d\bar{x}}{dt} = \bar{f}(t, \bar{x}, \mu),$$ (2.80)

$$\bar{x}(t_0) = \bar{x}_0,$$ (2.81)

where $\bar{x} \in \mathbb{R}^n$, $\bar{f} \in C(\mathbb{R}_+ \times \mathbb{R}^n \times M, \mathbb{R}^n)$.

It is assumed that its solution $\bar{x}(t) = \bar{x}(t, t_0, \bar{x}_0, \mu)$ exists for all $t \geq t_0$.

The system of equations (2.80) is a correct approximation of the system of equations (2.76) if the properties of the solutions of the system (2.80) are "close" to those of the solutions of the system (2.76) in a certain sense.

Remark 2.2. The system of equations (2.80) can be built on the basis of the system (2.76) in various ways; for example, by using some smoothing operator (see [46, 84]), by discarding a part of the equation (2.23), or when some parameters of the system (2.76) tend to certain limits as $t \to \infty$ (see [34] pp. 369, 379).

Remark 2.3. The dash over the n vectors x and f in the system (2.80) does not mean that these variables are related to the averaging process of the system (2.76).

2.5.2   Estimation of the Lyapunov function

Let, for the Lyapunov function $V(t, x, \mu)$, $V(t, 0, \mu) = 0$, $V \in C(\mathbb{R}_+ \times \mathbb{R}^n \times M, \mathbb{R}_+)$, its total derivative $D^+V(t, x, \mu)$ due to the system (2.79) be calculated using the formula

$$D^+V(t, x, \mu) = \limsup \left\{ \left[V\left(t + \theta, x + \theta \left(\sum_{i=1}^{m} \mu^i f_i(t, x), \mu \right) \right) \right.\right.$$
$$\left.\left. - V(t, x, \mu) \right] \theta^{-1} : \theta \to 0^+ \right\},$$

where $0 < m < \infty$ is an arbitrarily large number.

Let us show that the following assertion holds.

Lemma 2.4. *Suppose that for the system* (2.76), *there exists a function* $V(t, x, \mu)$ *with the properties indicated above. There exist, integrable on* $J \subseteq \mathbb{R}_+$, *non-negative functions* $\psi_i(t, \mu)$, $i = 1, 2, \ldots, m$, *such that*

$$D^+V(t, x, \mu) \leq \sum_{i=1}^{m} \psi_i(t, \mu) V^i(t, x, \mu) \qquad (2.82)$$

for all $(t, x) \in J \times D$.

Then, along the solutions of the system (2.79), *the variation of the function* $V(t, x, \mu)$ *is estimated using the inequality*

$$V(t, x(t), \mu) \leq V(t_0, x_0, \mu) \exp \left(\int_{t_0}^{t} \psi_1(s, \mu) ds \right) [L^*(t_0, t)]^{-\frac{1}{m-1}} \qquad (2.83)$$

for all $t \in J^* \subseteq J$, *for which*

$$L^*(t_0, t) = 1 - (m - 1) \int_{t_0}^{t} \sum_{i=2}^{m} V^{i-1}(t_0, x_0, \mu) \psi_i(s, \mu)$$

$$\times \exp \left(\int_{t_0}^{s} (m - 1) \psi_1(\tau) d\tau \right) ds > 0. \qquad (2.84)$$

The *proof* of Lemma 2.4 is similar to that of Lemma 1.6 (see also [103]), and hence it is not given here.

2.5.3   Deviation estimates

Consider the systems of equations (2.76) and (2.80) and assume that:

B_1. for the initial conditions of the solutions of the systems (2.76) and (2.80), the condition $x_0 \neq \bar{x}_0$ is satisfied;

B_2. there is a positive integrable function $g(t, \mu)$ such that

$$\|f(t, x, \mu) - \bar{f}(t, \bar{x}, \mu)\| \le g(t, \mu)\|x - \bar{x}\|^p, \ p > 1$$

in the region D;

B_3. the solution $\bar{x}(t) = \bar{x}(t, t_0, \bar{x}_0, \mu)$ of the system (2.80) is defined and remains in D for all $t \ge t_0$.

Let us calculate the value $T(\mu)$ using the formula

$$T(\mu) = \max_t \min_\mu \{(t, \mu) \in \mathbb{R}_+ \times M : N^*(t, \mu) < 1\},$$

where

$$N^*(t, \mu) = (p - 1)\|x_0 - \bar{x}_0\|^{p-1} \int_{t_0}^t g(s, \mu)ds.$$

Let us show that the following assertion holds.

Lemma 2.5. *Let the systems of equations* (2.76) *and* (2.80) *satisfy the conditions of assumptions B_1-B_3 and, in addition,*

$$N(t, \mu) = 1 - (p - 1)\|x_0 - \bar{x}_0\|^{p-1} \int_{t_0}^t g(s, \mu)ds > 0 \qquad (2.85)$$

for all $t \in [t_0, T(\mu)]$, where $\mu < \mu^* \in M$.
Then, the norm $\|x(t) - \bar{x}(t)\|$ satisfies the estimate

$$\|x(t) - \bar{x}(t)\| \le \|x_0 - \bar{x}_0\|(N(t, \mu))^{-\frac{1}{p-1}} \qquad (2.86)$$

for all $t \in [t_0, \ T(\mu)]$ and $0 < \mu < \mu^*$.

Proof. From the fact that

$$x(t) = x_0 + \int_{t_0}^t f(s, x(s), \mu)ds$$

and

$$\bar{x}(t) = \bar{x}_0 + \int_{t_0}^t \bar{f}(s, \bar{x}(s), \mu)ds,$$

due to conditions B_1–B_3, we find the estimate

$$\|x(t) - \bar{x}(t)\| \le \|x_0 - \bar{x}_0\| + \int_{t_0}^t g(s, \mu)\|x(s) - \bar{x}(s)\|^p ds \qquad (2.87)$$

for all $t \in [t_0, \ T(\mu)]$. □

From the inequality (2.87), it follows that

$$\|x(t) - \bar{x}(t)\| \leq \|x_0 - \bar{x}_0\| + \int_{t_0}^{t} g(s, \mu)\|x(s) - \bar{x}(s)\|^{p-1}\|x(s) - \bar{x}(s)\|ds.$$

$$(2.88)$$

Since $\|x_0 - \bar{x}_0\| > 0$, Lemma 1.2 applies to (2.88), which leads to the inequality

$$\|x(t) - \bar{x}(t)\| \leq \|x_0 - \bar{x}_0\| \exp\left(\int_{t_0}^{t} g(s, \mu)\|x(s) - \bar{x}(s)\|^{p-1}ds\right) \quad (2.89)$$

for all $t \in [t_0, \ T(\mu)]$. To evaluate the expression

$$\exp\left(\int_{t_0}^{t} g(s, \mu)\|x(s) - \bar{x}(s)\|^{p-1}ds\right),$$

transform the inequality (2.89) into the form

$$\|x(s) - \bar{x}(s)\|^{p-1} \leq \|x_0 - \bar{x}_0\|^{p-1}$$

$$\times \exp\left((p-1)\int_{t_0}^{t} g(s, \mu)\|x(s) - \bar{x}(s)\|^{p-1}ds\right). \quad (2.90)$$

Multiplying both sides of the inequality (2.90) by the negative factor $-(p-1)g(t, \mu)$, $0 < \mu < \mu^*$, we get

$$-(p-1)g(t, \mu)\|x(t) - \bar{x}(t)\|^{p-1} \geq -(p-1)\|x_0 - \bar{x}_0\|^{p-1}g(t, \mu)$$

$$\times \exp\left((p-1)\int_{t_0}^{t} g(s, \mu)\|x(s) - \bar{x}(s)\|^{p-1}ds\right) \quad (2.91)$$

for all $t \in [t_0, \ T(\mu)]$.

The inequality (2.91) is equivalent to the following:

$$-(p-1)g(t, \mu)\|x(t) - \bar{x}(t)\|^{p-1}$$

$$\times \exp\left((p-1)\int_{t_0}^{t} g(s, \mu)\|x(s) - \bar{x}(s)\|^{p-1}ds\right)$$

$$\geq -(p-1)\|x_0 - \bar{x}_0\|^{p-1}g(t, \mu) \quad (2.92)$$

for all $t \in [t_0, \ T(\mu)]$.

It is easy to see that

$$\frac{d}{dt}\left\{\exp\left(-(p-1)\int_{t_0}^{t} g(s, \mu)\|x(s) - \bar{x}(s)\|^{p-1}ds\right)\right.$$

$$\geq -(p-1)\|(x_0 - \bar{x}_0\|^{p-1}g(t, \mu)$$

and, further,

$$\exp\left(-(p-1)\int_{t_0}^{t} g(s,\mu)\|x(s)-\bar{x}(s)\|^{p-1}ds\right)$$

$$\geq 1-(p-1)\|x_0-\bar{x}_0\|^{p-1}\int_{t_0}^{t} g(s,\mu)ds.$$

Under the condition $N(t,\mu) > 0$ for all $t \in [t_0, \ T(\mu)]$, we have the estimate

$$\exp\left(-(p-1)\int_{t_0}^{t} g(s,\mu)\|x(s)-\bar{x}(s)\|^{p-1}ds\right) \leq (N(t,\mu))^{-1} \qquad (2.93)$$

for all $t \in [t_0, \ T(\mu)]$.

The inequalities (2.90) and (2.93) imply the estimate

$$\|x(t)-\bar{x}(t)\|^{p-1}\|x_0-\bar{x}_0\|^{-(p-1)} \leq (N(t,\mu))^{-1} \qquad (2.94)$$

for all $t \in [t_0, \ T(\mu)]$, from which we obtain the assertion of Lemma 2.5.

2.5.4   Estimating the norm of the difference between solutions

For the system of equations (2.79), we construct an *averaged system* of equations,

$$\frac{d\xi}{dt} = \mu\bar{f}_1(\xi)+\cdots+\mu^m\bar{f}_m(\xi), \qquad (2.95)$$

$$\xi(t_0) = \xi_0, \qquad (2.96)$$

by changing variables (see [39, p. 151])

$$x = \xi + \mu u_1(t,\xi) + \mu^2 u_2(t,\xi) + \cdots, \qquad (2.97)$$

where $u_k(t,\xi)$, $k = 1, 2, \ldots m$, are some functions to be defined.

We make the following assumptions about the systems of equations (2.79) and (2.95):

B_4. The vector functions $f_k(t,x) \in C(\mathbb{R}_+ \times \mathbb{R}^n, \ \mathbb{R}^n)$ are defined and bounded in t in the domain $D^* \subseteq (\mathbb{R}_+ \times \mathbb{R}^n)$, together with their even derivatives up to the $(m-1)$-th order.

B_5. There are limits at each point of D^*

$$\bar{f}_k(x) = \lim_{T\to\infty} \frac{1}{T}\int_{t_0}^{T} f_k(s,x)ds,$$

for all $k = 1, 2, \ldots, m$.

B_6. The solution $\xi(t)$, $\xi(t_0) \in D$, of the initial problem (2.95) and (2.96) is defined for all $t \geq t_0$ and is in the region D^*.

B_7. There are positive integrable functions $\psi_k(t, \mu)$, $k = 1, 2, \ldots, m$ such that

$$\|\mu^k(f_k(t, x) - \bar{f}_k(\xi))\| \leq \psi_k(t, \mu)\|x - \xi\|^k, \ k = 1, 2, \ldots, m,$$

for all $(t, x, \xi) \in D$.

B_8. For the initial values x_0 and ξ_0 of the solutions $x(t)$ and $\xi(t)$ of the systems (2.79) and (2.95), the condition $x_0 \neq \xi_0$ is satisfied.

Under conditions B_4–B_8, we estimate the norm of the *difference of solutions* $\|x(t) - \xi(t)\|$ on some interval of t.

Calculate the value of $T(\mu)$ using the formula

$$T(\mu) = \max_t \min_\mu \{(t, \mu) \in \mathbb{R}_+ \times M : \Phi^*(t, \mu) < 1\},$$

where

$$\Phi^*(t, \mu) = (m - 1) \int_{t_0}^t \sum_{k=2}^m \|x_0 - \xi_0\|^{k-1} \psi_i(s, \mu)$$

$$\times \exp\left(\int_{t_0}^t (m - 1)\psi_1(\tau, \mu)d\tau\right) ds,$$

and consider solutions of the initial problems (2.77)–(2.79) and (2.95)–(2.96) on the interval $[t_0, \ T(\mu)]$ for $0 < \mu < \mu^*$.

Lemma 2.6. *Let the systems of equations (2.79) and (2.95) be defined in the domain D and conditions B_4–B_8 be satisfied and, in addition,*

$$\Phi^*(t, \mu) = (m - 1) \int_{t_0}^t \sum_{k=2}^m \psi_k(s, \mu)\|x_0 - \xi_0\|^{k-1}$$

$$\times \exp\left((m - 1) \int_{t_0}^t \psi_1(\tau, \mu)d\tau\right) ds < 1$$

for all $t \in [t_0, \ T(\mu)]$.

Then, the norm of the solution difference $\|x(t) - \xi(t)\|$ satisfies the estimate

$$\|x(t) - \xi(t)\| \leq \|x_0 - \xi_0\| \exp\left(\int_{t_0}^t \psi_1(s, \mu)ds\right)(1 - \Phi^*(t, \mu))^{-\frac{1}{m-1}}$$

$$(2.98)$$

for all $t \in [t_0, \ T(\mu)]$, $0 < \mu < \mu^*$.

Proof.   Recall that the functions $\bar{f}_k(\xi)$ and $u_k(t, \xi)$, $k = 1, 2, \ldots, m$, are calculated using the formulas (see [39, p. 152])

$$\bar{f}_k(\xi) = \lim_{T \to \infty} \frac{1}{T} \int_0^T [f_k(s, \xi) + F_k(s, \xi)] ds,$$

$$u_k(t, \xi) = \int_0^t [f_k(s, \xi) + F_k(s, \xi) - \bar{f}_k(\xi(s))] ds + \varphi_k(\xi(t)), \ \ k = 1, 2, \ldots, m,$$

where the functions $\varphi_k(\xi(t)), k = 1, 2, \ldots, m$ are arbitrary; in particular, $\varphi_k(\xi(t)) = 0$, $k = 1, 2, \ldots, m$, $F_1(t, \xi) = 0$.

$$F_2(t, \xi) = \frac{\partial f_1}{\partial \xi} u_1(t, \xi) - \frac{\partial u_1}{\partial \xi} \bar{f}_1(\xi),$$

functions of $F_k(t, \xi)$ are known functions of u_1, $u_2, \ldots u_{k-1}$, $f_1(t, x), \ldots f_{k-1}(t, x)$, and their partial derivatives up to order $k - 1$. $\square$

Thus, the system of averaged equations (2.95) is completely determined by the system of equations (2.79). From the systems of equations (2.79) and (2.95), we obtain

$$\|x(t) - \xi(t)\| \le \|x_0 - \xi_0\| + \int_{t_0}^t \left(\sum_{k=1}^m \mu^k f_k(s, x(s)) - \sum_{k=1}^m \mu^k \bar{f}_k(\xi(s)) \right) ds$$
$$(2.99)$$

for all $t \in [t_0, \ T(\mu)]$. From the inequality (2.99), under condition B_7, we obtain the estimate

$$\|x(t) - \xi(t)\| \le \|x_0 - \xi_0\| + \int_{t_0}^t \sum_{k=2}^m \psi_k(s, \mu) \|x(s) - \xi(s)\|^m ds. \quad (2.100)$$

Since, according to assumption B_8, the condition $\|x_0 - \xi_0\| > 0$ is satisfied, Lemma 1.2 can be applied to the inequality (2.100), as a result of which we obtain the estimate

$$\|x(t) - \xi(t)\| \le \|x_0 - \xi_0\|$$
$$\times \exp \left[\int_{t_0}^t \left(\psi_1(s, \mu) + \sum_{k=2}^m \psi_k(s, \mu) \|x(s) - \xi(s)\|^{m-1} \|x(s) - \xi(s)\| \right) \right] ds.$$
$$(2.101)$$

Applying to the inequality (2.101) the estimation technique from the proof of Lemma 2.5, we obtain

$$\|x(t) - \xi(t)\| \le \|x_0 - \xi_0\| \exp\left(\int_{t_0}^{t} \sum_{k=2}^{m} \psi_1(s, \mu)ds\right)$$

$$\times \left\{ 1 - (m-1) \int_{t_0}^{t} \sum_{k=2}^{m} \psi_k(s, \mu)\|x_0 - \xi_0\|^{k-1} \right.$$

$$\left. \times \exp\left[(m-1)\int_{t_0}^{t} \psi_1(\tau, \mu)d\tau\right] ds \right\}^{\frac{1}{m-1}} \qquad (2.102)$$

for all $t \in [t_0, \ T(\mu)]$.

This proves Lemma 2.6.

2.5.5   Applications

Let us apply the estimates obtained in Lemmas 2.4–2.6 to solve the problem of stability on a finite interval and estimate the approximate integrations of systems with an asymptotic expansion of the right-hand side of the systems of equations.

2.5.5.1   *Conditions of $(\lambda, A, t_0, T(\mu))$-stability of motion*

An analysis of the *stability* of the system (2.76) with the expansion (2.79) on a *finite time interval* leads to $(\lambda, A, t_0, T(\mu))$-stability if the values λ, A, t_0, and $T(\mu)$ are fixed (see [34, 81]).

Definition 2.10. Given the estimates $0 < \lambda < A$, $t_0 \in \mathbb{R}_+$, and $T(\mu) > 0$ for $0 < \mu < \mu^*$, the motion of the system (2.79) is $(\lambda, A, t_0, T(\mu))$-stable if under the initial conditions $\|x(t_0)\| < \lambda$, the condition $\|x(t, \mu)\| < A$ is satisfied for all $t \in [t_0, \ T(\mu)]$.

The estimate (2.83) allows one to obtain sufficient conditions for the $(\lambda, A, t_0, T(\mu))$-stability of the system (2.79) in the following form.

Theorem 2.14. *For the system* (2.79), *let the function* $V(t, x, \mu)$ *be constructed in some way in the form of a positive-definite quadratic form. Let c_1 be the exact maximum of the function $V(t, x)$ on the sphere $\|x\| = \lambda$ and c_2 be the exact lower limit of $V(t, x)$ on the sphere $\|x\| = A$.*

If the conditions of Lemma 2.4 are satisfied and, in addition:

(1) *there exists* $0 < \mu_1 \in M$ *such that*

$$\Phi(t, c_1) = (m - 1) \int_{t_0}^{t} \sum_{i=2}^{m} c_1^{n-1} \psi_1(s, \mu)$$

$$\times \exp\left(\int_{t_0}^{s} (m - 1)\psi_1(\tau, \mu)d\tau \right) ds < 1$$

for $0 < \mu < \mu_1$ *and for all* $t \in [t_0, T(\mu)]$;

(2) *there exists a value* $0 < \mu_2 \in M$ *such that*

$$\exp\left(\int_{t_0}^{s} \psi_1(s, \mu)ds \right) (1 - \Phi(t, c_1))^{-\frac{1}{m-1}} < \frac{c_2}{c_1}$$

for $0 < \mu < \mu_2$, *then the motion of the system* (2.76) *with the decomposition of the right-hand side* (2.78) *is* $(\lambda, A, t_0, T(\mu))$-*stable, where* $0 < \mu < \min(\mu_1, \mu_2) \in M$.

Proof. Lemma 2.4 implies an estimate for the function $V(t, x, \mu)$ in the form (2.83) for which $V(t_0, x_0, \mu) \leq c_1$. Under conditions (1) and (2) of Theorem 2.14, we have the inequality $V(t, x(t), \mu) < c_2$ for all $t \in [t_0, T(\mu)]$ and, therefore, $\|x(t, \mu)\| < A$. This proves Theorem 2.14. $\square$

2.5.5.2 *ε-estimate for the norm of the difference of solutions* $\|x(t) - \bar{x}(t)\|$

The effective use of the approximating equations (2.80) in applied problems is related to the possibility of obtaining an ε-*estimate* for the *norm* of the difference of solutions $\|x(t) - \bar{x}(t)\|$ for all $t \geq t_0$ or for $t \in [t_0, T(\mu)]$, $T(\mu) > 0$.

Definition 2.11. Let us state that the approximating system (2.80) admits an ε-estimate of the approximate solution of the system of equations (2.76) if, under the initial conditions $x_0 \neq \bar{x}_0$, for all $t \geq t_0$ or $t \in [t_0, T(\mu)]$, the inequality $\|x(t) - \bar{x}(t)\| < \varepsilon$, where $x(t)$ is the solution to the initial problem (2.76)–(2.77) and $\bar{x}(t)$ is the solution to the initial problem (2.80)–(2.81).

The following assertion holds.

Theorem 2.15. *Assume that the systems of perturbed motion equations* (2.76) *and* (2.80) *satisfy the conditions of Lemma 2.5 and, in addition, for an arbitrarily small* $\varepsilon > 0$,

$$(N(t, \mu))^{-\frac{1}{p-1}} < \frac{\varepsilon}{\|x_0 - \bar{x}_0\|} \tag{2.103}$$

for all $t \geq t_0$ *or for* $t \in [t_0, T(\mu)]$.

Then, the norm $\|x(t) - \bar{x}(t)\|$ satisfies the ε-estimate, i.e., $\|x(t) - \bar{x}(t)\|$
$< \varepsilon$ for all $t \geq t_0$ or for $t \in [t_0, \ T(\mu)]$.

Proof.   The assertion of Theorem 2.15 follows from the estimate (2.97)
when the inequality (2.103) holds.
    The results obtained in this section may be of interest in problems of
qualitative analysis of the motion of affine systems and control systems of
general form (see [11] and the bibliography therein).
    The estimates (2.86) and (2.98) allow one to solve some problems of
interest in the theory of nonlinear oscillations and celestial mechanics (see
[46] and the bibliography therein).
    These problems include the following: the problem of conditions under
which for an arbitrary $\varepsilon > 0$ with the initial conditions $x_0 \neq \xi_0$, for solutions
$x(t)$ and $\xi(t)$, the inequality $\|x(t) - \xi(t)\| < \varepsilon$ is satisfied on the maximum
time interval; the problem of calculating $\sup \|x(t) - \xi(t)\|$ for $t \in [t_0, t_0 + T]$
and given $T > t_0 \in \mathbb{R}_+$; and the problem of determining the maximum
time interval T on which the inequality $\|x(t) - \xi(t)\| < K$ is satisfied for a
given value $0 < K < \infty$. $\square$

2.6   Analysis of Essentially Nonlinear Systems

Here, the general concept of a non-autonomous essentially nonlinear system
is introduced, and research problems are formulated.
    The boundaries of Lyapunov functions on solutions of essentially
nonlinear systems are established.
    Some classes of essentially nonlinear systems are considered, and condi-
tions are established for β-boundedness of solutions, (ε, δ, J)-stability in one
critical case; (ε, δ, J)-stability in the presence of an analytic integral, and
(ε, δ, J)-stability in a second-order system with slowly varying parameters.

2.6.1   Statement of the problem

Consider the system of equations of perturbed motion

$$\frac{dx}{dt} = A(t)x + X(t, x), X(t, x) = \sum_{j=1}^{m} X^{(j)}(t, x), \qquad (2.104)$$

where $x \in \mathbb{R}^n$, $A(t)$ is an $n \times n$ matrix with continuous elements on $\mathbb{R}_+ =$
$[0, \infty)$ and $X^{(j)}(t, x)$ is a vector of functions vanishing at $x_1 = \cdots = x_n = 0$

and continuous in $(t, x) \in \mathbb{R}_+ \times D \backslash \{0\}$, where $D \subset \mathbb{R}^n$ is an open area in $\mathbb{R}^n$.

Definition 2.12. The system of equations (2.104) is essentially nonlinear if $A(t) \equiv 0$ and $x \in D \backslash \{0\}$, the vector functions $X^{(j)}(t, x)$ satisfy the condition

$$\|X^{(j)}(t, x)\| \le k_j(t) \|x\|^{p_j}, \qquad (2.105)$$

where $k_j(t)$ are non-negative functions on $\mathbb{R}_+$ and $1 < p_j < \infty$ for all $j = 1, 2, \ldots, m$. Here, $\| \cdot \|$ is the Euclidean norm of the vector x in the space $\mathbb{R}^n$.

It is of interest to analyse the stability and equi-boundedness of solutions of non-autonomous *essentially nonlinear systems* of equations based on the development of the direct Lyapunov method in combination with nonlinear integral inequalities.

2.6.2   Bound of variation of Lyapunov functions

Suppose that for the system (2.104), a function $V : [0, \infty) \times \mathbb{R}^n \to \mathbb{R}_+$ is constructed and is differentiable with respect to $(t, x) \in \mathbb{R}_+ \times D \backslash \{0\}$ and definitely positive. The total derivative of the function $V(t, x)$ on the solutions of the system (2.104) is calculated using the formula

$$V'(t, x(t)) = \frac{\partial V}{\partial t}(t, x) + \left(\frac{\partial V}{\partial x}(t, x)\right)^T (A(t)x + X(t, x)). \qquad (2.106)$$

Let us indicate the boundary of the variation of the function $V(t, x)$ on the solutions of the system (2.105) under the initial conditions of motion

$$x(t_0) = x_0, \quad x_0 \in D \backslash \{0\}. \qquad (2.107)$$

The following assertion holds.

Lemma 2.7. *If the system (2.105) has a function $V(t, x)$ and continuous non-negative functions $\psi_i(t)$, $i = 1, 2, \ldots m$, $m > 1$, such that:*

(1) $\frac{\partial V}{\partial t}(t, x) + \left(\frac{\partial V}{\partial x}(t, x)\right)^T A(t)x \le 0$ *for all* $(t, x) \in \mathbb{R}_+ \times D \backslash \{0\}$,

(2) $\left(\frac{\partial V}{\partial x}(t, x)\right)^T \left(\sum_{j=1}^{m} X^{(j)}(t, x)\right) \le \sum_{j=1}^{m} \psi_i(t) V^{(j)}(t, x)$ *for all* $(t, x) \in$ $\mathbb{R}_+ \times D \backslash \{0\}$,

then $V(t, x(t))$ satisfies the estimate

$$V(t, x(t)) \le V(t_0, x_0) \exp\left(\int_{t_0}^{t} \psi_1(s) ds\right)\left(1 - M(t, t_0)\right)^{-\frac{1}{m-1}} \qquad (2.108)$$

for those values $t \in J \subset \mathbb{R}_+$ for which $t \geq t_0$,

$$M(t, t_0) = (m-1) \int_{t_0}^{t} \sum_{j=2}^{m} V^{(j-1)}(t_0, x_0) \psi_j(s)$$

$$\times \exp\left(\int_{t_0}^{s} (m-1)\psi_1(\tau) ds \right) ds < 1. \qquad (2.109)$$

Proof. From the relation (2.106) and the conditions of Lemma 2.7, we obtain the inequality

$$V'(t, x(t)) \leq \sum_{j=1}^{m} \psi_j(t) V^{(j)}(t, x(t)) \qquad (2.110)$$

for all $(t, x) \in \mathbb{R}_+ \times D \setminus \{0\}$. From the inequality (2.110), it follows that

$$V(t, x(t)) \leq V(t_0, x_0) + \int_{t_0}^{t} \sum_{j=1}^{m} \psi_j(s) V^{(j)}(s, x(s)) ds$$

$$= V(t_0, x_0) + \int_{t_0}^{t} \left(\psi_1(s) + \sum_{j=2}^{m} \psi_j(s) V^{(j-1)}(s, x(s)) \right) V(s, x(s)) ds.$$

$$(2.111)$$

Applying to the inequality (2.111) the technique of proving Theorem 2.6 from [72], we obtain the estimate (2.108). This proves Lemma 2.7. ☐

Remark 2.4. Here and in the following, to estimate the interval $J = [t_0, T)$, we use a formula of the form

$$T \leq \sup\{ t \in \mathbb{R}_+ : M(t, t_0) < 1 \}.$$

Next, consider an essentially nonlinear system, (2.105), in which

$$X(t, x) = X^{(1)}(t, x) + X^{(2)}(t, x), \qquad (2.112)$$

where $X^{(j)}(t, x)$, $j = 1, 2$, satisfy the conditions (2.106). Let us show that the following assertion holds.

Lemma 2.8. *Assume for the system (2.105), there exist a function $V(t, x)$ and continuous non-negative functions $a_1(t)$ and $a_2(t)$ such that:*

(1) *condition (1) of Lemma 2.7 is satisfied;*
(2)

$$\left(\frac{\partial V}{\partial x}(t, x) \right)^T \left(\sum_{j=1}^{2} X^{(j)}(t, x) \right) \leq a_1(t) V^p(t, x) + a_2(t, x) V^q(t, x)$$

$$(2.113)$$

for all $(t,x) \in J \times D \backslash \{0\}$, $1 < p < q < \infty$ and for all $t \in J$, for which

$$N(t, t_0) = (p + q - 2)$$

$$\times \left(V^{p-1}(t_0, x_0) \int_{t_0}^{t} a_1(s)ds + V^{q-1}(t_0, x_0) \int_{t_0}^{t} a_2(s)ds \right) < 1.$$

Then, the function $V(t, x(t))$ satisfies the estimate

$$V(t, x(t)) \leq V(t_0, x_0)(1 - N(t, t_0))^{-\frac{1}{p+q-2}} \qquad (2.114)$$

for all $t \in J$.

Proof. Conditions (1) and (2) of Lemma 2.8 imply that

$$V(t, x(t)) \leq V(t_0, x_0) + \int_{t_0}^{t} (a_1(s)V^p(s, x(s)) + a_2(s)V^q(s, x(s)))ds$$

$$= V(t_0, x_0) + \int_{t_0}^{t} (a_1(s)V^{p-1}(s, x(s)) + a_2(s)V^{q-1}(s, x(s)))V(s, x(s))ds.$$

$$(2.115)$$

□

Further, applying the technique of estimating from Lemma 1.7 for the inequality (2.115), we obtain the estimate (2.114). This proves Lemma 2.8.

Example 2.6. Consider the equations of perturbed motion of the form

$$\frac{dx}{dt} = m(t)x + y + g(t)x(x^2 + y^2),$$

$$\frac{dy}{dt} = m(t)y - x + g(t)y(x^2 + y^2),$$

$$(2.116)$$

where x, $y \in \mathbb{R}$ and $m(t)$ and $g(t)$ are some continuous non-negative functions for all $t \in \mathbb{R}_+$.

For the function $V'(x, y) = x^2 + y^2$, the total derivative along the solutions of the system (2.116) has the form

$$D^+V(x, y) = 2m(t)V(x, y) + 2g(t)V^2(x, y)$$

for all $(x, y) \in \mathbb{R} \times \mathbb{R}$ and $t \in \mathbb{R}_+$.

Suppose that

$$N_1(t, t_0) = 2 \left(\int_{t_0}^{t} m(s)ds + (x_0^2 + y_0^2) \int_{t_0}^{t} g(s)ds \right) < 1$$

for all $t \in J$. Then, according to Lemma 2.8

$$x^2(t) + y^2(t) \le (x_0^2 + y_0^2)(1 - N_1(t, t_0))^{-1} \qquad (2.117)$$

for all $t \in J$.

If for any $0 < \varepsilon < H$, we choose $\delta = \delta(\varepsilon) > 0$ so that $x_0^2 + y_0^2 < \delta$, then under the condition

$$(1 - N_1(t, t_0))^{-1} < \frac{\varepsilon}{\delta} \quad \text{for all } t \in J,$$

we obtain an estimate for the solutions of the system (2.116) in the form $x^2(t) + y^2(t) < \varepsilon$ for all $t \in J$.

Remark 2.5. Here, to estimate the interval $J = [t_0, T)$, we use the formula

$$T \le \sup \left\{ t \in \mathbb{R}_+ : N_1(t, t_0) < \frac{1}{2} \right\}.$$

2.6.3 Applications

The estimates for the Lyapunov function presented in Lemmas 2.7 and 2.8 make it possible to study a wide range of problems in the qualitative analysis of essentially nonlinear systems. Some of these problems are discussed in this section.

2.6.3.1 *β-boundedness of motions of a second-order system*

Consider a *weakly perturbed system* of equations

$$\begin{aligned} \frac{dx}{dt} &= X_1(t, x, y) + \mu X_2(t, x, y), \\ \frac{dy}{dt} &= Y_1(t, x, y) + \mu Y_2(t, x, y), \end{aligned} \qquad (2.118)$$

where $X_i \in C(\mathbb{R}_+ \times \mathbb{R} \times \mathbb{R}, \ \mathbb{R})$, $Y_i \in C(\mathbb{R}_+ \times \mathbb{R} \times \mathbb{R}, \ \mathbb{R})$, $i = 1, 2$, and $\mu \in [0, 1)$ is a small parameter. According to [55], the system (2.118) is essentially nonlinear if it does not become linear for $\mu = 0$. For the system (2.118), we consider the function

$$V(x, y) = x^2 + y^2 \qquad (2.119)$$

and assume that there are continuous non-negative functions $\bar{c}_1(t, \mu)$ and $\bar{c}_2(t, \mu)$ such that the inequalities

$$\begin{aligned} 2x(X_1(t, x, y) + \mu X_2(t, x, y)) &\le \bar{c}_1(t, \mu)(x^2 + y^2)^2, \\ 2y(Y_1(t, x, y) + \mu Y_2(t, x, y)) &\le \bar{c}_2(t, \mu)(x^2 + y^2)^3 \end{aligned} \qquad (2.120)$$

hold for all $(t, x, y) \in \mathbb{R}_+ \times D \backslash \{0\}$ and $\mu < \mu_1 \in [0, 1)$. It is easy to check that the function $V(x, y)$ on solutions of the system (2.119) satisfies the estimate

$$V(x(t, \mu),\ y(t, \mu)) \le V(x_0, y_0) \left(1 - N_2(t_0, t, \mu) \right)^{-\frac{1}{3}} \tag{2.121}$$

if only

$$N_2(t_0, t, \mu) = 3 \left(V(x_0, y_0) \int_{t_0}^{t} \bar{c}_1(s, \mu) ds + V^2(x_0, y_0) \int_{t_0}^{t} \bar{c}_2(s, \mu) ds \right) < 1$$

for all $t \in J$ and $\mu < \mu_2 \in [0, 1)$.

Further, the properties of the solutions are determined, taking into account the boundedness of the interval on which the behaviour of the solutions is considered, as well as the dependence of the solutions on the values of a small parameter.

Definition 2.13. The solution $(x(t, \mu),\ y(t, \mu))^T$ of the system (2.118) is *β-bounded* if for a given $\beta > 0$ and some $\mu^* \in [0, 1)$,

$$x^2(t, \mu) + y^2(t, \mu) < \beta \quad \text{for all} \quad t \in J$$

and $\mu < \mu^*$, where β may depend on each solution of the system (2.118).

Definition 2.14. A solution $(x(t, \mu), y(t, \mu))^T$ of the system (2.118) is *(α, β)-bounded* if for a given $\beta > 0$, for any $\alpha > 0$ and $t_0 \in \mathbb{R}_+$, there exists $\mu^* \in [0, 1)$ such that if $x_0^2 + y_0^2 < \alpha^2$, then

$$x^2(t, \mu) + y^2(t, \mu) < \beta$$

for all $t \in J$ and $\mu < \mu^*$.

The following assertion holds.

Theorem 2.16. *Let the system* (2.118) *satisfy the conditions* (2.120) *and* $N_2(t, t_0, \mu) < 1$ *for all* $t \in J$ *and* $\mu < \min(\mu_1, \mu_2)$.
If for any $(x_0, y_0) \in \mathbb{R}$, *the following inequality is fulfilled:*

$$\left(1 - N_2(t, t_0, \mu) \right)^{-\frac{1}{3}} < \frac{\beta}{x_0^2 + y_0^2} \quad \text{for all} \quad t \in J \tag{2.122}$$

and $\mu < \min(\mu_1, \mu_2)$, $\beta > 0$, *then the solution of the system* (2.118) *is β-bounded.*

Proof. It follows from the inequality (2.121) and the bound (2.122) that $x^2(t, \mu) + y^2(t, \mu) < \beta$ for all $t \in J$ and $\mu < \min(\mu_1, \mu_2)$. $\qquad \square$

Remark 2.6. Here, to estimate the interval $J = [t_0, T)$, we use the formula

$$T \leq \sup\{t \in \mathbb{R}_+ : N_2(t, t_0, \mu) < 1\}.$$

Theorem 2.17. *Let the system* (2.118) *satisfy the condition* (2.120) *and*

$$N_2^*(t, t_0, \mu) = 3\frac{\alpha^2}{2}\left(\int_{t_0}^t \bar{c}_1(s, \mu)ds + \frac{\alpha^2}{2}\int_{t_0}^t \bar{c}_2(s, \mu)ds\right) < 1$$

for all $t \in J$ *and* $\mu < \mu_2^* \in [0, 1)$.
    If the initial values $(x_0, y_0) \in \mathbb{R}$ *are such that* $x_0^2 + y_0^2 \leq \alpha^2$ *and*

$$\left(1 - N_2^*(t, t_0, \mu)\right)^{-\frac{1}{3}} < \frac{\beta(t_0, \alpha)}{\alpha^2} \quad \text{for all} \quad t \in J$$

and $\mu < \min(\mu_1, \mu_2^*)$, *then the solution of the system* (2.118) *is* (α, β)-*bounded.*

Proof.   The proof of Theorem 2.18 is similar to the proof of Theorem 2.17.   □

Remark 2.7. Here, to estimate the interval $J = [t_0, T)$, we use the formula

$$T \leq \sup\{t \in \mathbb{R}_+ : N_2^*(t, t_0, \mu) < 1\}.$$

This proves Theorem 2.17.

2.6.3.2   *Critical case of r pairs of pure imaginary roots in the absence of resonance*

Consider the non-autonomous system of equations

$$\frac{dx_s}{dt} = -\lambda_s y_s + \sum_{j=m}^N X_s^{(j)}(t, x, y),$$

$$\frac{dy_s}{dt} = \lambda_s x_s + \sum_{j=m}^N Y_s^{(j)}(t, x, y), \tag{2.123}$$

$$s = 1, 2, \ldots, r.$$

Here, $\lambda_s > 0$, $x \in \mathbb{R}^r$, $y \in \mathbb{R}^r$, $X_s^{(j)}(t, x, y)$, and $Y_s^{(j)}(t, x, y)$ are homogeneous forms in x and y of order $j(j \geq 2)$, continuous in $t \in \mathbb{R}_+$ and vanishing at $x = y = 0$.
    For the total derivative of the function

$$V(x_s, y_s) = \sum_{s=1}^r (x_s^2 + y_s^2) \tag{2.124}$$

along the solutions of the system (2.123), it is easy to obtain the expression

$$\frac{dV}{dt}(x_s, y_s) = \sum_{s=1}^{r} \left(2x_s \sum_{j=m}^{N} X_s^{(j)}(t, x, y) + 2y_s \sum_{j=m}^{N} Y_s^{(j)}(t, x, y) \right) \quad (2.125)$$

for all $t \in \mathbb{R}_+$ and $(x, y) \in D^* \setminus \{0\}$, $D^* \subseteq \mathbb{R}^r \times \mathbb{R}^r$.

Suppose that there are continuous non-negative functions $\bar{a}_1(t)$, $\bar{a}_2(t)$ and constant $1 < p < q < +\infty$ such that the right-hand side of the relation (2.125) is evaluated by an inequality

$$\sum_{s=1}^{r} \left(2x_s \sum_{j=m}^{N} X_s^{(j)}(t, x, y) + 2y_s \sum_{j=m}^{m} Y_s^{(j)}(t, x, y) \right)$$

$$\leq \bar{a}_1(t) V^p(x_s, y_s) + \bar{a}_2(t) V^q(x_s, y_s) \quad (2.126)$$

and the condition

$$N_3(t, t_0) = (p + q - 2) \left(V^{p-1}(x_{s_0}, y_{s_0}) \int_{t_0}^{t} \bar{a}_1(s) ds \right.$$

$$\left. + V^{q-1}(x_{s_0}, y_{s_0}) \int_{t_0}^{t} \bar{a}_2(s) ds \right) < 1 \quad (2.127)$$

holds for all $t \in J$.

When the inequalities (2.126) and (2.127) are satisfied, according to Lemma 2.8, for the function (2.124) on the solutions of the system (2.123), we obtain the estimate

$$V(x_s(t), y_s(t)) \leq V(x_{s_0}, y_{s_0})(1 - N_3(t, t_0))^{-\frac{1}{p+q-2}} \quad (2.128)$$

for all $t \geq t_0$.

Definition 2.15. A solution $(x(t), y(t))^T = 0$ of the system (2.123) is (ε, δ, J)-*stable* if for any $\varepsilon > 0$ and $t_0 \in \mathbb{R}_+$, there exists $\delta(t_0, \varepsilon) > 0$ such that if $\sum_{s=1}^{r}(x_{s,0}^2 + y_{s,0}^2) < \delta$, then

$$\sum_{s=1}^{r}(x_s^2(t) + y_s^2(t)) < \varepsilon$$

for all $t \in J$.

The following assertion holds.

Theorem 2.18. *Let the system of equations (2.123) and the Lyapunov function (2.124) satisfy the conditions (2.126) and (2.127) and, moreover,* *for any $\varepsilon > 0$, there exists $\delta(\varepsilon) > 0$ such that for $\sum_{s=1}^{r}(x_{s_0}^2 + y_{s_0}^2) < \delta(\varepsilon)$, the inequality*

$$(1 - N_3(t, t_0))^{-\frac{1}{p+q-2}} < \frac{\varepsilon}{\delta(\varepsilon)} \quad \text{holds for all} \quad t \in J. \tag{2.129}$$

Then, the zero solution of the system (2.123) is (ε, δ, J)-stable.

Proof.   The assertion of Theorem 2.18 follows from the estimates (2.129) and (2.128) since when they are satisfied, we obtain that $\sum_{s=1}^{r}(x_s^2(t, \mu) + y_s^2(t, \mu)) < \varepsilon$ for all $t \in J$. $\square$

Remark 2.8. Here, to estimate the interval $J = [t_0, T)$, we use the formula

$$T \leq \sup\{t \in \mathbb{R}_+ : N_3(t, t_0) < 1\}.$$

2.6.3.3   *(ε, δ, J)-stability of solutions to a system with a sign-definite integral*

An essentially nonlinear non-autonomous second-order system is considered:

$$\begin{aligned}
\frac{dx}{dt} &= -\lambda y + X_0(t, x, y) + \mu X_1(t, x, y) + \cdots, \\
\frac{dy}{dt} &= \lambda x + Y_0(t, x, y) + \mu Y_1(t, x, y) + \cdots.
\end{aligned} \tag{2.130}$$

Here, $x, y \in \mathbb{R}$, $\lambda > 0$, and $\mu \in [0, 1)$ is a small parameter. The functions X_0 and Y_0 are analytic in the variables x and y, continuous in $t \in \mathbb{R}_+$, and do not contain linear terms. The functions X_i and Y_i, $i = 1, 2, \ldots$, are polynomials in x and y, continuous in $t \in \mathbb{R}_+$, and $X_i(t, 0, 0) = Y_i(t, 0, 0) = 0$ for all $t \geq t_0$.

Suppose that for $\mu = 0$, the *system* (2.130) admits the existence of an *analytic integral*

$$x^2 + y^2 + \mu F(t, x, y) \tag{2.131}$$

for all $(t, x, y) \in \mathbb{R}_+ \times D \setminus \{0\}$, where $F(t, x, y)$ is a function continuous in $t \in \mathbb{R}_+$ with bounded partial derivatives with respect to all variables. We take the Lyapunov function $V(t, x, y)$ in the form

$$V(t, x, y) = x^2 + y^2 + \mu F(t, x, y). \tag{2.132}$$

Given that (2.130) is an analytic integral of the system (2.132) with $\mu = 0$, for the total derivative of the function (2.132) along the solutions of the system (2.130), we have the expression

$$\frac{dV}{dt}(t, x, y) = 2\mu \left(x \sum_{i=1}^{\infty} X_i(t, x, y) + y \sum_{i=1}^{\infty} Y_i(t, x, y) \right)$$

$$+ \mu \left[\left(\frac{\partial F}{\partial x} \right) \sum_{i=1}^{\infty} X_i(t, x, y) + \left(\frac{\partial F}{\partial y} \right) \sum_{i=1}^{\infty} Y_i(t, x, y) \right].$$

$$(2.133)$$

Let there exist continuous non-negative functions $b_1(t, \mu)$ and $b_2(t, \mu)$ and constants $1 < p < q < \infty$ such that

$$2\mu \left(x \sum_{i=1}^{\infty} X_i(t, x, y) + y \sum_{i=1}^{\infty} Y_i(t, x, y) \right) \le b_1(t, \mu) V^p(t, x, y), \qquad (2.134)$$

$$\mu \left[\left(\frac{\partial F}{\partial x} \right) \sum_{i=1}^{\infty} X_i(t, x, y) + \left(\frac{\partial F}{\partial y} \right) \sum_{i=1}^{\infty} Y_i(t, x, y) \right] \le b_2(t, \mu) V^q(t, x, y)$$

$$(2.135)$$

for all $(t, x, y) \in \mathbb{R}_+ \times D \setminus \{0\}$ and $\mu < \mu_1 \in [0, 1)$.

Suppose also that there exists $\mu_2 \in [0, 1)$ such that

$$N(t, t_0, \mu) = (p + q - 2) \left(V^{p-1}(t_0, x_0, y_0) \int_{t_0}^{t} b_1(s, \mu) ds \right.$$

$$\left. + V^{q-1}(t_0, x_0, y_0) \int_{t_0}^{t} b_2(s, \mu) ds \right) < 1 \qquad (2.136)$$

for all $t \in J$ and for $\mu < \mu_2$.

Under the conditions (2.134) and (2.135), for the function $V(t, x, y)$, it is easy to obtain an estimate of the form

$$V(t, x(t), y(t)) \le V(t_0, x_0, y_0)(1 - N(t, t_0, \mu))^{-\frac{1}{p+q-2}} \qquad (2.137)$$

for all $t \in J$ and $\mu < \min(\mu_1, \mu_2)$.

Remark 2.9. Here, to estimate the interval $J = [t_0, T)$, we use the formula

$$T \le \sup\{t \in \mathbb{R}_+ : N(t, t_0, \mu) < 1\}.$$

Next, based on the estimate (2.130) for the function $V(t, x, y)$ on the solutions of the system (2.137), we establish conditions for the uniform (α, β, J)-stability of zero solutions of the system (2.130).

Theorem 2.19. *Assume that for the system* (2.130), *the Lyapunov function* (2.132) *is constructed, for which the following conditions are satisfied:*

(1) *there are comparison functions a and b belonging to K-class such that*

$$a(\|x\| + \|y\|) \leq V(t, x, y) \leq b(\|x\| + \|y\|)$$

for all $(t, x, y) \in \mathbb{R}_+ \times D \setminus \{0\}$;

(2) *the relations* (2.134), (2.135), *and* (2.136) *hold for all* $t \in J$ *and* $\mu < \min(\mu_1, \mu_2)$;

(3) *there exist a constant* $0 < k < \infty$ *such that*

$$(1 - N(t, t_0, \mu))^{-\frac{1}{p+q-2}} \leq k$$

uniformly in $t_0 \in \mathbb{R}_+$ *for* $\mu < \min(\mu_1, \mu_2)$.

Then, the solution $x(t, \mu) = y(t, \mu) = 0$ *of the system* (2.130) *is uniformly* (ε, δ, J)-*stable.*

Proof.   By virtue of condition (1) of Theorem 2.19, for any $0 < \varepsilon < H$, the estimate $a(\varepsilon) \leq V(t, x, y)$ is true for all $t \in J$ and $(x, y) \in D \setminus \{0\}$ such that $\|x\| + \|y\| = \varepsilon$. Given $\varepsilon > 0$, choose $\delta = \delta(\varepsilon) > 0$ so that for $b(\delta)$ is true inequality $\|x_0\| + \|y_0\| < b(\delta)k < a(\varepsilon)$. Further, under condition (2) of Theorem 2.19, we have the estimate (2.137) for the function $V(t, x, y)$. Let $\|x(t_1)\| + \|y(t_1)\| = \varepsilon$ hold for some $t_1 > t_0$, $(t_0, t_1) \in J$. From the estimate (2.137), under condition (3) of Theorem 2.19, we have a contradiction

$$a(\varepsilon) \leq V(t_1, x(t_1), y(t_1)) \leq V(t_0, x_0, y_0)k \leq b(\delta)k < a(\varepsilon),$$

which implies that $\|x(t)\| + \|y(t)\| < \varepsilon$ for all $t \in J$ uniformly in $t_0 \in J$ for $\mu < \min(\mu_1, \mu_2)$. This proves Theorem 2.19. □

2.6.3.4 (ε, δ, J)-*stability in a system with slowly varying parameters*

Consider the system of equations of perturbed motion

$$\frac{dx}{dt} = -\lambda(\tau)y + X_0(x, y) + \sum_{i=1}^{\infty} \mu^i X_i(\tau, x, y),$$

$$\frac{dy}{dt} = \lambda(\tau)x + Y_0(x, y) + \sum_{i=1}^{\infty} \mu^i Y_i(\tau, x, y). \qquad (2.138)$$

Here, $x, y \in \mathbb{R}$, $\lambda(\tau)$ is a bounded, continuous, and single-valued function of τ, $\tau = \mu t$ is *slow time*, X_i and Y_i are slowly varying bounded functions of time, the series in μ is absolutely convergent in the range $(x, y) \in \mathbb{R}^2$ for small values of the parameter μ, and X_0 and Y_0 are nonlinear functions of the variables x and y. The functions $X_i = 0$, $Y_i = 0$ and $X_0 = 0$, $Y_0 = 0$ for $x = y = 0$ and for all $\tau \in [0, \infty)$.

It is assumed that the roots of the characteristic equation of the system (2.138) have no intersection points for all $\tau \in [0, \infty)$, and for $\mu = 0$, the system of equations (2.138) is *neutral stable*. In particular, this will happen if

$$xX_0(x, y) + yY_0(x, y) \leq 0 \qquad (2.139)$$

for all $(x, y) \in D \setminus \{0\}$.

With the system of equations (2.138), we consider the Lyapunov function $V(x, y) = \frac{1}{2}(x^2 + y^2)$ in the range $(x, y) \in D \setminus \{0\}$.

Under (2.139), for the total derivative of the function $V(x, y)$ on the solutions of the system (2.138), we have the expression

$$\frac{dV}{dt}(x, y) \leq x \left(\sum_{i=1}^{\infty} \mu^i X_i(\tau, x, y) \right) + y \left(\sum_{i=1}^{\infty} \mu^i Y_i(\tau, x, y) \right) \qquad (2.140)$$

for all $(x, y) \in \mathbb{R}$ and $\tau \in \mathbb{R}_+$.

Let there exist $\mu_1 \in [0, 1)$ and $1 < q < \infty$ and a continuous non-negative function $\sigma(\tau, \mu)$ such that

$$x \left(\sum_{i=1}^{\infty} \mu^i X_i(\tau, x, y) \right) + y \left(\sum_{i=1}^{\infty} \mu^i Y_i(\tau, x, y) \right) \leq \sigma(\tau, \mu) V^q(x, y)$$

$$(2.141)$$

for all $(x, y) \in D \setminus \{0\}$, $\mu < \mu_1$ and the inequality

$$\Phi(\tau, \tau_0, \mu) = (q - 1)V^{q-1}(x_0, y_0) \int_{\tau_0}^{\tau} \sigma(s, \mu)ds < 1 \qquad (2.142)$$

holds for all $\mu < \mu_2 \in [0, 1)$ and $\tau \in J$.

It is easy to show that when the inequalities (2.141) and (2.142) hold, the function $V(x, y)$ has the estimate

$$V(x(\tau, \mu), y(\tau, \mu)) \leq V(x_0, y_0)(1 - \Phi(\tau, \tau_0, \mu))^{-\frac{1}{q-1}} \qquad (2.143)$$

for all $\mu < \min(\mu_1, \mu_2)$ and $\tau \in J$.

Remark 2.10. Here, to estimate the interval $J = [t_0, T)$, we use the formula

$$T \leq \sup\{t \in \mathbb{R}_+ : \Phi(\tau, \tau_0, \mu) < 1\}.$$

The estimate of the function $V(x, y)$ in the form (2.143) allows us to establish the following result.

Theorem 2.20. *Assume that the system* (2.138) *satisfies the following conditions*:

(1) *for* $\mu = 0$, *the system is neutrally stable*;
(2) *for* $\mu < \mu_1$, *there exists a function* $\sigma(\tau, \mu)$ *for which the inequality* (2.141) *is satisfied*;
(3) *for* $\mu < \mu_2$, *the inequality* (2.143) *holds for all* $\tau \in J$;
(4) *for any* $\varepsilon > 0$, *there exists* $\delta = \delta(\tau_0, \varepsilon) > 0$ *such that for* $x_0^2 + y_0^2 \leq \delta(\tau_0, \varepsilon)$, *the inequality*

$$(1 - \Phi(\tau, \tau_0, \mu))^{-\frac{1}{q-1}} < \frac{\varepsilon}{\delta}$$

holds for all $\tau \in J$ *and* $\mu < \min(\mu_1, \mu_2)$.

Then, the zero solution of the essentially nonlinear system (2.138) *is* (ε, δ, J)-*stable*.

Proof. Under the conditions of Theorem 2.20, for the function $2V(x, y) = x^2 + y^2$, we have the estimate (see Lemma 2.5)

$$x^2(\tau) + y^2(\tau) \leq \delta(1 - \Phi(\tau, \tau_0, \mu))^{-\frac{1}{q-1}}$$

for all $\tau \in J$ and $\mu < \min(\mu_1, \mu_2)$, which implies that $x^2(\tau) + y^2(\tau) < \varepsilon$ for all $\tau \in J$.

The conditions for β-boundedness and (ε, δ, J)-stability of solutions of essentially nonlinear systems obtained in this section are of interest for further development and applications in control problems for the motion of a rigid body in the theory of nonlinear oscillations, as well as in the theory of complex systems. □

2.7 Comments and Bibliography

The key element of Lyapunov's [76] function method is the estimation of the variation (decrease or increase) in the auxiliary function along the solutions of the equations of perturbed motion. One of the directions of development of the Lyapunov function method is to obtain more refined estimates of the total derivative of the Lyapunov function on solutions of the corresponding equations of perturbed motion.

In fact, when studying specific equations of perturbed motion (linear or nonlinear), the problem of estimating the variation of the Lyapunov function is reduced to constructing such a majorant of the total derivative of this function, at which it is possible to draw a conclusion about the decrease or increase in Lyapunov functions over time (see Burton [26], Moiseev [120], and the bibliography therein).

This chapter is devoted to one of the possible ways of solving this problem based on the use of nonlinear integral inequalities.

Section 2.1 adapts the results from the paper by Martynyuk, Stamova I., and Stamov G. [94]. In addition, it provides a brief analysis of the estimates of the Lyapunov function based on the differential inequalities; see Aleksandrov and Platonov [4], which were used in the works of Lyapunov [76], Brauer [24], and Melnikov [119].

Section 2.2 is based on the results of the paper by Martynyuk, Khusainov, and Chernienko [107, 108].

Section 2.3 uses the results of the paper by Martynyuk [91].

Note that among the mathematical models of real systems and processes encountered in engineering and technology, differential equations containing a small parameter occupy a special place (see Bogoliubov and Zubarev [23], Grebenikov and Riabov [46], Krylov and Bogoliubov [63], Martynyuk [84], Martynyuk *et al.* [101], Poincaré [137], and the bibliography therein).

Section 2.4 considers systems with an asymptotic expansion of the right-hand side of a system of differential equations following the paper by Martynyuk and Chernienko [103].

Section 2.5 is based on the results of the development presented in the paper by Martynyuk and Chernienko [106].

In the work of Veretennikov [166], essentially nonlinear systems are defined as such systems of differential equations of perturbed motion that do not become linear as a small parameter tends to zero. Such systems allow the use of the Kamenkov [55] method of analysing oscillations if, at μ equal to zero, the system has one or a family of periodic solutions.

Chapter 3

Stability and Stabilisation of Motion of Polynomial and Interval Systems

3.1 Introduction

This chapter is devoted to the dynamic analysis of motion of polynomial systems of differential equations and of systems of equations with interval initial conditions. Namely, estimates of the variation of Lyapunov functions on solutions of polynomial systems are established, on the basis of which sufficient conditions for various types of motion stability are established. Here, the conditions for stabilisation of motion to various types of stability are also obtained. The chapter ends with the stability analysis of the zero solution of a polynomial system with aftereffect.

The chapter is organised according to the following plan.

Section 3.2 formulates the problem of estimating the Lyapunov function for a polynomial system of differential equations. On the basis of the estimates obtained, the results of the study of stability by Lyapunov, practical stability, and stability on a finite interval under large initial perturbations are presented.

In Section 3.3, we study the stability of a polynomial system with aftereffect based on Lyapunov functionals and functions.

Section 3.4 presents a method for stabilising solutions of a polynomial system based on the method of Lyapunov functions.

Section 3.5 presents one approach to the problem of stabilising the motion of nonlinear systems under interval initial conditions.

The final Section 3.6 contains comments on the problem under discussion and a bibliography of works on this topic.

3.2 Stability Analysis of a Polynomial System

This section presents the results of the analysis of Lyapunov stability, practical stability, and stability on a finite interval of the zero solution of a *polynomial system* based on new estimates for the variation of the Lyapunov function on solutions of the system of equations of perturbed motion under consideration.

3.2.1 Some definitions and problem statement

Consider the equations of perturbed motion of some mechanical system in the form (see [6] and the bibliography therein)

$$\frac{dx_\beta}{dt} = \sum_{s=2h-1}^{2l-1} X_\beta^{(s)}(x, a_{i_1 i_2, \ldots i_s}^{(\beta)}(t)), \tag{3.1}$$

$$x_\beta(t_0) = x_{\beta 0}, \quad \beta = 1, 2, \ldots, n, \tag{3.2}$$

where $x_\beta \in \mathbb{R}$, $x \in \mathbb{R}^n$, $a_{i_1, i_2, \ldots, i_s}^{(\beta)}(t)$ are the real-time functions bounded on any finite interval, $x_1, \ldots, x_n$ are the deviations of the variables $x_\beta(t)$ from the state $x(t) = 0$ for any $t \in \mathbb{R}_+$, $0 < h \le l$ are positive integers, and

$$X_\beta^{(s)}(x, a_{i_1 i_2, \ldots, i_s}^{(\beta)}(t)) = \sum_{i_1=1}^{n} \sum_{i_2=i_1}^{n} \cdots \sum_{i_s=i_{s-1}}^{n} a_{i_1 i_2, \ldots, i_s}^{(\beta)}(t) x_{i_1} x_{i_2} \cdots x_{i_s}$$

$$\tag{3.3}$$

is a homogeneous polynomial in n variables $x_1, x_2, \ldots, x_n$ with reduced similar terms arranged in *lexicographic order*.

Recall that the lexicographic order of $x_1 > x_2 > \cdots > x_n$ is given by the rule

$$x_1^{k_1} \cdots x_n^{k_n} > x_1^{l_1} \cdots x_n^{l_n} \iff (\exists i : k_i > l_i \text{ and } k_j = l_j \text{ when } j < i).$$

Let us suppose that for the system of equations (3.1), a *polynomial function* $V(t, x) > 0$, is constructed in some way for all $x \in \mathbb{R}^n \backslash 0$ and $t \in \mathbb{R}_+$.

Recall that a function $V(t, x) > 0$, $V(t, 0) = 0$, is called a *Lyapunov polynomial function* if $V(t, x)$ is a polynomial of some even degree and, together with the total derivative by virtue of the system (3.1), solves the problem of stability (instability) of the state $x(t) = 0$ of the system (3.1).

Some examples of polynomial Lyapunov functions are given in [6, 155] and other works, where some of their properties are discussed.

Next, we consider the problem of determining the boundary of the variation of the function $V(t,x)$ along the solutions of the system (3.1) for all $t \in \mathbb{R}_+$.

3.2.2 Estimation of the Lyapunov function

For the total time derivative of the function $V(t,x)$, due to the system (3.1),

$$V'(t,x) = \frac{\partial V}{\partial t} + \sum_{\beta=1}^{n} \frac{\partial V}{\partial x_\beta} \frac{dx_\beta}{dt}$$

assume the existence of continuous non-negative functions $\psi_i(t)$, $i = 1, 2, \ldots, m$, such that

$$V'(t,x) \leq \sum_{i=1}^{m} \psi_i(t) V^i(t,x) \tag{3.4}$$

for all $(t,x) \in \mathbb{R}_+ \times \mathbb{R}^n$. It follows from the inequality (3.4) that the function $V(t,x) > 0$ satisfies the inequality

$$V(t,x(t)) \leq V(t_0,x_0) + \int_{t_0}^{t} \sum_{i=1}^{m} \psi_i(s) V^i(s,x(s)) ds \tag{3.5}$$

for all $t \in \mathbb{R}_+$. The following assertion holds.

Lemma 3.1. *Let the system (3.1) have a definitely positive function $V(t,x)$ and non-negative functions $\psi_i(t)$ integrable on $\mathbb{R}_+$, $i = 1, 2, \ldots, m$, such that for all $(t,x) \in \mathbb{R}_+ \times \mathbb{R}^n$, the estimate (3.4) is fulfilled.*

Then, the boundary of the variation of the function $V(t,x)$ along the solutions of the polynomial system (3.1) is estimated by the inequality

$$V(t,x(t)) \leq V(t_0,x_0) \exp\left(\int_{t_0}^{t} \psi_1(s) ds \right) \left(1 - N(t_0,t) \right)^{-\frac{1}{m-1}} \tag{3.6}$$

for all $t \in \mathbb{R}_+$, for which

$$N(t_0,t) = (m-1) \int_{t_0}^{t} \sum_{i=2}^{m} V^{i-1}(t_0,x_0) \psi_i(s) \exp\left(\int_{t_0}^{s} (m-1)\psi_1(\tau) d\tau \right) ds < 1. \tag{3.7}$$

Proof. We rewrite the estimate (3.5) as

$$V(t, x(t)) \leq V(t_0, x_0) + \int_{t_0}^t \left(\psi_1(s)V(s, x(s)) + \sum_{i=2}^m \psi_i(s)V^i(s, x(s)) \right) ds$$

(3.8)

and, further,

$$V(t, x(t)) \leq V(t_0, x_0) + \int_{t_0}^t \left(\psi_1(s) + \sum_{i=2}^m \psi_i(s)V^{i-1}(s, x(s)) \right) V(s, x(s)) ds.$$

(3.9)

□

Since $V(t_0, x_0) > 0$, the Gronwall–Bellman lemma applies to (3.9). As a result, we obtain the estimate

$$V(t, x(t)) \leq V(t_0, x_0) \exp \left[\int_{t_0}^t \left(\psi_1(s) + \sum_{i=2}^m \psi_i(s)V^{i-1}(s, x(s)) \right) ds \right]$$

(3.10)

for all $(t, x) \in \mathbb{R}_+ \times \mathbb{R}^n$. To evaluate the expression

$$\exp \left[\int_{t_0}^t \left(\sum_{i=2}^m \psi_i(s)V^{i-1}(s, x(s)) \right) ds \right]$$

in the inequality (3.10), we apply a modification of the approach from [72, 82]. For any $i = 2, \ldots, m$, from the inequality (3.10), we find

$$V^{i-1}(t, x(t))$$
$$\leq V^{i-1}(t_0, x_0) \exp \left[(i-1) \int_{t_0}^t \left(\psi_1(s) + \sum_{i=2}^m \psi_i(s)V^{i-1}(s, x(s)) \right) ds \right]$$
$$\leq V^{i-1}(t_0, x_0) \exp \left[(m-1) \int_{t_0}^t \left(\psi_1(s) + \sum_{i=2}^m \psi_i(s)V^{i-1}(s, x(s)) \right) ds \right].$$

(3.11)

Multiplying both sides of the inequality (3.11) by the negative factor $-(m-1)\,\psi_i(t)$, we obtain the estimate

$$-(m-1)\psi_i(t)V^{i-1}(t, x(t)) \exp \left[-(m-1) \int_{t_0}^t \sum_{i=2}^m \psi_i(s)V^{i-1}(s, x(s)) ds \right]$$
$$\geq -(m-1)\psi_i(t)V^{i-1}(t_0, x_0) \exp \left[(m-1) \int_{t_0}^t \psi_1(s) ds \right].$$

(3.12)

Summing up both parts of the inequality (3.12) over i from 2 to m, we obtain the inequality

$$-(m-1)\sum_{i=2}^{m}\psi_i(t)V^{i-1}(t,x(t))$$

$$\times \exp\left[-(m-1)\int_{t_0}^{t}\sum_{i=2}^{m}\psi_i(s)V^{i-1}(s,x(s))ds\right]$$

$$\geq -(m-1)\sum_{i=2}^{m}\psi_i(t)V^{i-1}(t_0,x_0)\exp\left[(m-1)\int_{t_0}^{t}\psi_1(s)ds\right].$$

$$(3.13)$$

From the inequality (3.13), it follows that

$$\exp\left(-(m-1)\int_{t_0}^{t}\sum_{i=2}^{m}\psi_i(s)V^{i-1}(s,x(s))ds\right)$$

$$\geq 1-(m-1)\int_{t_0}^{t}\sum_{i=2}^{m}\psi_i(s)V^{i-1}(t_0,x_0)\exp\left((m-1)\int_{t_0}^{s}\psi_1(\tau)d\tau\right)ds.$$

$$(3.14)$$

From the inequality (3.14), we find that

$$\exp\left(\int_{t_0}^{t}\sum_{i=2}^{m}\psi_i(s)V^{i-1}(s,x(s))ds\right)$$

$$\leq \left(1-N(t_0,t)\right)^{-\frac{1}{m-1}}.$$

$$(3.15)$$

Given the estimate (3.15), the inequality (3.10) becomes

$$V(t,x(t))$$

$$\leq V(t_0,x_0)\exp\left(\int_{t_0}^{t}\psi_1(s)ds\right)\left(1-N(t_0,t)\right)^{-\frac{1}{m-1}}$$

for all $t \in \mathbb{R}_+$, for which the inequality (3.7) holds. This proves Lemma 3.1.

Corollary 3.1. *Let the estimate* (3.9) *be a function* $\psi_1(s) = 0$ *for all* $t \in \mathbb{R}_+$. *Then, the estimate* (3.6) *becomes*

$$V(t,x(t)) \leq \frac{V(t_0,x_0)}{\left(1-(m-1)\int_{t_0}^{t}\sum_{i=2}^{m}\psi_i(s)V^{i-1}(t_0,x_0)ds\right)^{\frac{1}{m-1}}} \qquad (3.16)$$

for all $t \in \mathbb{R}_+$, for which

$$N^*(t_0, t) = (m-1) \int_{t_0}^t \sum_{i=2}^m \psi_i(s) V^{i-1}(t_0, x_0) ds < 1.$$

Corollary 3.2. *Let the estimate* (3.9) *of the functions* $\psi_i(t) = 0$, $i = 2, \ldots, m$, *hold for all* $t \in \mathbb{R}_+$. *Then, the estimate* (3.6) *becomes*

$$V(t, x(t)) \le V(t_0, x_0) \exp \left(\int_{t_0}^t \psi_1(s) ds \right)$$

for all $t \in \mathbb{R}_+$.

Together with the estimate (3.6), Corollaries 3.1 and 3.2 allow us to consider various problems of qualitative analysis of the motion of polynomial systems of the form (3.1).

Example 3.1. Let the perturbed motion of some mechanical system be described by the equation

$$\frac{dx}{dt} = a_1(t)x + a_2(t)x^3 + a_3(t)x^5, \quad x(t_0) = x_0, \tag{3.17}$$

where $x \in \mathbb{R}$, $a_i(t) \in C(\mathbb{R}, \mathbb{R}_+)$, $i = 1, 2, 3$.

We use the positive-definite function $V(x) = x^2$, $x \in \mathbb{R}$, and find an estimate for the function $V(x(t))$ on the solution $x(t)$ of the problem (3.17) based on Lemma 3.1. The derivative of this function due to the equation (3.17) has the form

$$\frac{dV(x(t))}{dt} = \sum_{i=1}^3 b_i(t) V^i(x(t)), \tag{3.18}$$

where $b_1(t) = 2a_1(t)$, $b_2(t) = 2a_2(t)$, $b_3(t) = 2a_3(t)$.

Let us suppose that

$$N(t_0, t) = \int_{t_0}^t \sum_{i=2}^3 V^i(x_0) b_i(s) \exp \left(2 \int_{t_0}^s b_1(\tau) d\tau \right) ds < \frac{1}{2} \tag{3.19}$$

for all $t \in \mathbb{R}_+$.

According to the formula (3.6), on the solutions of the polynomial equation (3.17), we have the estimate

$$V(x(t)) \le V(x_0) \exp \left(\int_{t_0}^t b_1(s) ds \right) \left(1 - N(t_0, t) \right)^{-\frac{1}{2}}$$

for all $t \in \mathbb{R}_+$, for which the condition (3.19) is satisfied.

Remark 3.1. If the inequality (3.19) holds on some finite interval, then the function $V(x(t))$ is estimated on the same interval.

3.2.3 Applications

In this section, the obtained estimates for the Lyapunov functions are used in some problems in the theory of motion stability of polynomial systems.

3.2.3.1 *Lyapunov stability*

Recall the definition of *Lyapunov stability* (see [76]).

Definition 3.1. The state $x = 0$ of the system (3.1) is:

(a) stable if, given $\varepsilon \in (0, H)$ and $t_0 \in \mathbb{R}_+$, there exists $\delta = \delta(t_0, \varepsilon) > 0$ such that for all $x_0 \in B_\delta$, the inequality $\|x(t, t_0, x_0)\| < \varepsilon$ holds for all $t \in \mathbb{R}_+$.
(b) uniformly stable if δ in Definition 3.1(a) does not depend on t_0, i.e., $\delta = \delta(\varepsilon)$.

Here and in the following, $B_\delta = \{x \in \mathbb{R}^n : \|x\| < \delta\}$ is an open domain in $\mathbb{R}^n$.

Theorem 3.1. *Let a function $V(t, x)$ be constructed for the system (3.1) and there exist comparison functions $a, b \in K$-class such that:*

(1) $a(\|x\|) \leq V(t, x) \leq b(\|x\|)$ *for any $(t, x) \in \mathbb{R}_+ \times \Omega$;*
(2) *there are positive continuous functions $\psi_i(t), i = 1, 2, \ldots, m$ and the inequality (3.4) is fulfilled;*
(3) *all conditions of Lemma 3.1 are satisfied;*
(4) *for given $\varepsilon > 0$ and $t_0 \in \mathbb{R}_+$, there exists $\delta = \delta(t_0, \varepsilon) > 0$ such that*

$$\exp\left(\int_{t_0}^{t} \psi_1(s)ds\right)(1 - N(t_0, t))^{-\frac{1}{m-1}} < \frac{a(\varepsilon)}{b(\delta)}$$

for all $t \in \mathbb{R}_+$.

Then, the state $x = 0$ of the polynomial system (3.1) is Lyapunov stable.

Proof. Let $t_0 \in \mathbb{R}_+$ and $\varepsilon \in (0, H)$, $H = const > 0$, be given. It follows from the condition (3.1) of Theorem 3.1 that $V(t_0, x_0) \leq b(\|x_0\|)$ for all $x_0 \in B_\delta$. Under the conditions (3.4) and (3.7), the estimate (3.6) is true along the solution $x(t, t_0, x_0)$ for any $x_0 \in B_\delta$ and $t_0 \in \mathbb{R}_+$. It follows from the condition (3.1) of the theorem that

$$\|x(t, t_0, x_0)\| \leq a^{-1}(V(t, x(t, t_0, x_0))) \quad \text{for all} \quad t \in \mathbb{R}_+.$$

From the estimate (3.6), under the condition (4) of Theorem 3.1, we have

$$\|x(t, t_0, x_0)\| \leq a^{-1}(V(t, x(t, t_0, x_0))) < a^{-1}(a(\varepsilon)) = \varepsilon$$

for all $t \in \mathbb{R}_+$. Since $a \in K$-class, we have $\|x(t, t_0, x_0)\| < \varepsilon$ for all $t \in \mathbb{R}_+$.
□

Theorem 3.1 is proved.

3.2.3.2 *Practical stability*

Recall the definition of practical stability (see [67]).

Definition 3.2. Given estimates for (λ, A), $0 < \lambda < A$, the *polynomial system* (3.1) is *practically stable* if for $\|x_0\| < \lambda$, the estimate $\|x(t, t_0, x_0)\| < A$ holds for all $t \in \mathbb{R}_+$.

Definition 3.3. A *Lyapunov function* $V(t, x)$ is said to be *locally large* if, for a given estimate A, for any $0 < c < A$ and $t_0 \geq 0$, there exists $\Delta = \Delta(t_0, c) > 0$ such that outside the spheres $\sum_{s=1}^{n} x_s^2 = \Delta^2$, the inequality $V(t, x) > c$ is valid for all $t \geq t_0$.

Remark 3.2. For a correct analysis of practical stability, the Lyapunov functions must be such that the level surfaces $V(t, x) = c$ are closed.

Remark 3.3. Locally large Lyapunov functions occupy an intermediate position between Lyapunov functions [76] and radially unbounded functions (see [169]).

Remark 3.4. The absence of locally large conditions in those investigated for practical stability is justified only in the case of applying a definitely positive Lyapunov function in the form of a quadratic form, which is radially unbounded and, therefore, locally large.

Definition 3.4. If the function $b(t, s) \in C(\mathbb{R}_+^2, \mathbb{R}_+)$ and $b(t, s) \in K$-class for each t, then $b(t, s)$ is said to belong to the class CK.

Theorem 3.2. *Let a locally large Lyapunov function $V(t, x)$ be constructed for the system (3.1) and there exist comparison functions $a \in KR$-class and $b \in CK$-class such that:*

(1) $a(\|x\|) \le V(t,x) \le b(t,\|x\|)$ *for all* $(t,x) \in \mathbb{R}_+ \times \Omega$;
(2) *conditions* (2) *and* (3) *of Theorem* 3.1 *are satisfied*;
(3) *given estimates for* (λ, A), *the following inequality holds*:

$$\exp\left(\int_{t_0}^{t} \psi_1(s)ds\right)(1 - N(t_0,t))^{-\frac{1}{m-1}} < \frac{a(A)}{b(t_0,\lambda)}$$

for all $t \in \mathbb{R}_+$.

Then, the polynomial system (3.1) *is practically stable.*

Proof.   Let (λ, A) be given. From the continuity of $V(t,x)$ and the fact that $V(t,0) = 0$, it follows that

$$V(t_0,x_0) < b(t_0,\|x_0\|) \quad \text{at} \quad \|x_0\| < \lambda.$$

Condition (2) of Theorem 3.2 implies that, along the solution $x(t,t_0,x_0)$ for $\|x_0\| < \lambda$, the estimate (3.6) holds for the function $V(t,x(t))$ for all $t \in \mathbb{R}_+$. Taking into account condition (1) of Theorem 3.2, we obtain

$$\|x(t)\| \le a^{-1}(V(t,x(t))) < a^{-1}(a(A)) = A \qquad (3.20)$$

for all $t \in \mathbb{R}_+$. Since the function $a \in KR$-class, the estimate (3.20) completes the proof of Theorem 3.2.   □

3.2.3.3   *Stability on a finite interval*

The theory of motion stability on a finite interval is closely related to the theory of practical motion stability.

Definition 3.5. Given estimates for the quantities (c_1, c_2, t_0, T), the *polynomial system* (3.1) is *stable* on a finite interval with respect to a locally large function $V(t,x)$ if, from the condition $V(t_0,x_0) \le c_1$, it follows that $V(t,x(t)) < c_2$ and $(c_1 < c_2)$ for all $t \in [t_0, t_0 + T]$ for the solution $x(t) = x(t,t_0,x_0)$.

Remark 3.5. If in Definition 3.5 the constants $c_1 = c_2 = c$, where c is an arbitrarily small positive value and the interval $[t_0, t_0 + T]$ is unbounded, then Definition 3.5 becomes Definition 20 from the monograph [176, p. 157].

Theorem 3.3. *Let a locally large function* $V(t,x)$ *be constructed for the system* (3.1) *and, in addition:*

(1) *all conditions of Corollary* 3.1 *are satisfied*;

(2) *given estimates for* (c_1, c_2, t_0, T), *the following inequality holds:*

$$(1 - N^*(t_0, t))^{-\frac{1}{m-1}} < \frac{c_2}{c_1} \qquad (3.21)$$

for all $t \in [t_0, t_0 + T]$.

Then, the polynomial system (3.1) *is stable on a finite interval in the sense of Definition 3.5.*

Proof. Let the solution $x(t, t_0, x_0)$ of the system (3.1) leave the domain $V(t_0, x_0) \le c_1$ at $t = t_0$. According to the conditions of Corollary 3.1, the estimate (3.16) is true along the solution $x(t, t_0, x_0)$ of the system (3.1). It follows from condition (2) of Theorem 3.3 that for all $t \in [t_0, t_0 + T]$, the estimate

$$V(t, x(t, t_0, x_0)) < c_2$$

is satisfied. Then, the polynomial system (3.1) is stable on a finite interval in the sense of Definition 3.5. $\square$

Example 3.2. Let the system of differential equations

$$\frac{dx_1}{dt} = x_1 + (x_1^2 + 3x_1 x_2^2)(u(t) + p(t)), \qquad (3.22)$$

$$\frac{dx_2}{dt} = x_2 + (x_2^3 - x_1^2 x_2)(u(t) + p(t)), \quad x_1(t_0) = x_{10}, \ x_2(t_0) = x_{20} \qquad (3.23)$$

describe some *positional differential game* (see [62]). Here, $t \in J$, $J = [t_0, t_0 + T]$ and functions $u(t) : J \to [-\alpha, \alpha]$, $p(t) : J \to [-\beta, \beta]$, $\alpha, \beta > 0$, and $\alpha > \beta$ such that $u(t) + p(t) \ge 0$ for all $t \in J$.

For a positive definite function $V(x) = x_1^2 + x_2^2$, we have the relation

$$\frac{dV(x(t))}{dt} = 2x_1^2 \left[1 + (x_1^2 + 3x_2^2)(u(t) + p(t)) \right]$$

$$+ 2x_2^2 \left[1 + (x_2^2 - x_1^2)(u(t) + p(t)) \right]$$

$$= 2(V(x) + V^2(x)(u(t) + p(t))). \qquad (3.24)$$

It implies that

$$V(x(t)) = V(x_0) + \int_{t_0}^{t} (2V(x(s)) + 2(u(t) + p(t))V^2(x(s)))ds. \qquad (3.25)$$

According to Lemma 3.1, we get

$$V(x(t)) = V(x_0) \exp \left[\int_{t_0}^{t} \psi_1(\tau) d\tau \right] (1 - L(t_0, t))^{-1} \qquad (3.26)$$

for all $t \in \mathbb{R}_+$, for which

$$L(t_0, t) = V(x_0) \int_{t_0}^{t} \psi_2(s) \exp \left[\int_{t_0}^{s} \psi_1(\tau) d\tau \right] ds < 1, \qquad (3.27)$$

where $\psi_1(t) = 2$; $\psi_2(t) = 2(u(t) + p(t))$.

It follows from the relation (3.27) that a *positional differential game* (3.23)–(3.24) is *stable* in the sense of Definition 3.5 if for all $t \in [t_0, \, t_0 + T]$, the condition (3.27) is satisfied and for given (c_1, c_2, t_0, T),

$$(1 - L(t_0, t))^{-1} < \frac{c_2}{c_1}$$

for all $t \in [t_0, \, t_0 + T]$.

3.3 Stability of a Polynomial System with Aftereffect

The problem of stability of solutions to systems of differential equations with aftereffect has been studied in many monographs (see [26, 51, 56, 160] and the bibliography therein). The problem of stability of solutions of polynomial equations with aftereffect is little studied (see [127]).

In this section, we consider a polynomial system with a general aftereffect, and based on new estimates of the Lyapunov functional (function), we obtain sufficient conditions for the stability and uniform boundedness of solutions of the equations under consideration.

3.3.1 Statement of the problem

We consider a polynomial system of equations of perturbed motion with aftereffect in the form

$$\frac{dx}{dt} = f(t, x, x_t). \qquad (3.28)$$

Here, $x \in \mathbb{R}^n$, $x_t(s) = x(t + s)$ for $-\tau \leq s \leq 0$. Denote by $(C, \| \cdot \|)$ the Banach space of continuous functions $\varphi : [-\tau, 0] \to \mathbb{R}^n$, with the norm $\|\varphi\| = \max_{-\tau \leq s \leq 0} |\varphi(s)|$, where $|\cdot|$ is an appropriate norm in $\mathbb{R}^n$. Further, the norm $|x| = \left(\sum_{i=1}^{n} x_i^2 \right)^{1/2}$ is applied. For a given $H > 0$, the notation C_H denotes a subset of functions from the space $(C, \| \cdot \|)$ for which $\|\varphi\| < H$,

$\dfrac{dx}{dt}$ denotes the right derivative of the state vector x of the system (3.28) at time t if it exists and is finite. The vector function $f : \mathbb{R} \times \mathbb{R}^n \times C_H \to \mathbb{R}^n$ is a continuous polynomial function, mapping bounded sets to bounded sets. Moreover, for each $t_0 \in \mathbb{R}$ and each $\varphi \in C_H$, there exists at least one solution $x_t(t_0, \varphi)$ of the equation (3.28) such that $x_{t_0} = \varphi$ is defined on the interval $[t_0, t_0 + \alpha]$, and if there exists $H_1 > 0$ for which $|x(t, t_0, \varphi)| < H_1$, then $\alpha = \infty$.

The vector function $f(t, x, x_t)$ has as its components homogeneous polynomials in the variables $x_1, x_2, \ldots, x_n$; $x_{1_t}, x_{2_t}, \ldots, x_{n_t}$ with given similar members, which are arranged in lexicographic order.

3.3.2 The method of Lyapunov functionals

Let the functional $V : \mathbb{R}_+ \times C_H \to \mathbb{R}_+$ be constructed for the system (3.28). The total derivative $V'(t, x_t(t_0, \varphi))$ along the solutions of the system (3.28) is defined by the formula (see [169, p. 186])

$$V'(t, x_t(t_0, \varphi)) = \limsup \left\{ [V(t + \delta, x_{t+\delta}(t, \varphi)) - V(t, \varphi)]\delta^{-1} : \delta \to 0^+ \right\},$$
(3.29)

where $x(t_0, \varphi)$ is the solution of the system (3.28) for $x_{t_0}(t_0, \varphi) = \varphi$.

Definition 3.6. A continuous functional $V(t, \varphi) : \mathbb{R}_+ \times C_H \to \mathbb{R}_+$ is said to be positive definite if there exists a function $a \in K$-class for which

$$V(t, \varphi) \geq a(|\varphi(0)|), \quad \varphi(s) \in C_H.$$

Assume that there are continuous non-negative functions $\psi_i(t)$, $i = 1, 2, \ldots, m$, such that

$$V'(t, x_t(t_0, \varphi)) \leq \sum_{i=1}^{m} \psi_i(t) V^i(t, x_t)$$
(3.30)

for all $(t, \varphi) \in \mathbb{R}_+ \times C_{H_1}$, $0 < H_1 < H$.

From the Lyapunov relation for the functional $V(t, \varphi)$ and the estimate (3.30) of its total derivative, due to the polynomial system (3.28), it follows that

$$V(t, x_t(t_0, \varphi)) \leq V(t_0, \varphi) + \int_{t_0}^{t} \sum_{i=1}^{m} \psi_i(s) V^i(s, x_s(t_0, \varphi)) ds$$
(3.31)

for all $t \in \mathbb{R}_+$.

Lemma 3.2. *Assume that for the system of equations (3.28), a positive-definite functional $V(t, \varphi)$ is constructed and there exist integrable non-negative functions $\psi_i(t)$, $i = 1, 2, \ldots, m$, such that the condition (3.30) is satisfied. Then, the limit of variation of the functional $V(t, \varphi)$ on the solutions of the system (3.28) is estimated using the inequality*

$$V(t, x_t(t_0, \varphi)) \leq V(t_0, \varphi) \exp \left(\int_{t_0}^t \psi_1(s) ds \right) (1 - M(t_0, t))^{-\frac{1}{m-1}} \qquad (3.32)$$

for all $t \geq t_0$, for which

$$M(t_0, t) = (m-1) \int_{t_0}^t \sum_{i=2}^m (V^{i-1}(t_0, \varphi) \psi_2(s))$$

$$\times \exp \left(\int_{t_0}^s (m-1) \psi_1(\tau) d\tau \right) ds < 1.$$

The proof of the estimate (3.32) is carried out similarly to that of Lemma 3.1.

Corollary 3.3. *Let the estimate (3.30) hold, $m = 2$, and the existence conditions for the functional $V(t, \varphi)$ and the functions $\psi_i(t)$, $i = 1, 2$, indicated in Lemma 3.2 be satisfied. Then, the boundary of the variation of the functional $V(t, \varphi)$ on the solutions of the system (3.28) is estimated using the inequality*

$$V(t, x_t(t, \varphi)) \leq V(t_0, \varphi) \exp \left(\int_{t_0}^t \psi_1(s) ds \right) (1 - L(t_0, t))^{-1} \qquad (3.33)$$

for all $t \geq t_0$, for which

$$L(t_0, t) = V(t_0, \varphi) \int_{t_0}^t \psi_2(s) \exp \left(\int_{t_0}^s \psi_1(\tau) d\tau \right) ds < 1.$$

The proof of the estimate (3.33) is carried out similarly to that of Lemma 3.2 from Chapter 1.

3.3.2.1 Stability of solutions

Recall some definitions (see [169, p. 251]).

Definition 3.7. Let $x(t) = 0$ be a solution to the system (3.28).

(a) The zero solution of the system (3.28) is *stable* if for each $\varepsilon > 0$ and $t_1 \geq t_0$, there exists $\delta > 0$ such that the condition $\varphi \in C_H$, $\|\varphi\| < \delta$, implies the estimate $|x(t, t_1, \varphi)| < \varepsilon$ for all $t \geq t_1$.

(b) The zero solution of the system (3.28) is *uniformly stable* if it is stable
and the value $\delta > 0$ from definition (a) can be chosen independently of
$t_1 \geq t_0$.

Let us show that the following assertion holds.

Theorem 3.4. *Assume that for a polynomial system,* (3.28), *for some*
$H > 0$, *there exists a functional* $V(t, \varphi)$ *continuous in* (t, φ) *for* $t_0 \leq t < \infty$
and $\varphi \in C_H$ *and* $V(t, 0) = 0$. *In addition, the following conditions are*
satisfied:

(1) *there is an angular function* $w : [0, \infty) \rightarrow [0, \infty)$, $w(0) = 0$, $w(r) > 0$
for $r > 0$, *strictly increasing and such that*

$$w(|\varphi(t)|) \leq V(t, \varphi)$$

for all $t_0 < t < \infty$ *and* $\varphi \in C_H$;
(2) *all conditions of Lemma 3.2 are satisfied and there exists a constant*
$0 < k < \infty$ *such that*

$$\exp\left(\int_{t_0}^t \psi_1(s)ds\right)(1 - M(t_0, t))^{-\frac{1}{m-1}} \leq k < \infty \qquad (3.34)$$

for all $t \geq t_0$.

Then, the zero solution of the polynomial system (3.28) *is stable.*

Proof. Let $t_1 \geq t_0$ and $\varepsilon > 0$ be given, $\varepsilon < H$. The continuity of the
functional $V(t, \varphi)$ and the fact that $V(t, 0) = 0$ imply the existence of
$\delta > 0$ such that for $\varphi \in C_H$, for $t = t_1$, $\|\varphi_{t_1}\| < \delta$, we have the estimate
$kV(t_0, \varphi) < w(\varepsilon)$. It follows from conditions (1) and (2) of Theorem 3.4
that

$$w(|x(t, t_1, \varphi_{t_1})|) \leq V(t, x_t(t_1, \varphi_{t_1})) \leq kV(t_0, \varphi) < w(\varepsilon).$$

This implies that $|x(t, t_1, \varphi_{t_1})| < \varepsilon$ for all $t \geq t_1$. This proves
Theorem 3.4. □

Theorem 3.5. *If, in addition to conditions* (1) *and* (2) *of Theorem 3.4,*
there exists an angular function $w_1(r)$ *such that* $V(t, \varphi) \leq w_1(\|\varphi\|)$ *and the*
estimate (3.34) *is uniform in* $t_0 \in \mathbb{R}_+$, *then the zero solution of the system*
(3.28) *is uniformly stable.*

Proof.   Let $\varepsilon > 0$, $\varepsilon < H$ be given, and $\delta > 0$ be found such that $w_1(\delta) < w(\varepsilon)$. If $t_1 \geq t_0$ and $\varphi \in C_H$, $\|\varphi_{t_1}\| < \delta$, then we have the inequalities

$$w(|x(t, t_1, \varphi_{t_1})|) \leq V(t, x_t(t_1, \varphi_{t_1})) \leq kV(t_0, \varphi)$$

$$\leq w_1(\|\varphi_{t_1}\|) < w_1(\delta) < w(\varepsilon).$$

Hence, it follows that $|x(t, t_1, \varphi_{t_1})| < \varepsilon$ uniformly in $t_1 \in \mathbb{R}_+$. This proves Theorem 3.5.   □

3.3.2.2   *Uniform boundedness of solutions*

In this section, we assume that the polynomial system (3.28) is defined and continuous in the range $(t, \varphi) \in \mathbb{R}_+ \times C$, where C is the space of functions mapping $[-h, 0] \to \mathbb{R}^n$. Let S^* be a subset of functions $\varphi \in C$, for which $|\varphi(0)| \geq H$, where H can be large. Let $f(t, x, x_t) \neq 0$ for all $t \geq t_0$ and $x = x_t = 0$.

Definition 3.8. The solutions $x(t, t_0, \varphi)$ of the polynomial system (3.28) are:

(a) *equi-bounded* if for any $\alpha > 0$ and $t_0 \in \mathbb{R}_+$, there exists $\beta(t_0, \alpha) > 0$ such that if $\varphi \in S^*$, $\|\varphi\| < \alpha$, then $|x(t, t_0, \varphi)| < \beta(t_0, \alpha)$ for all $t \geq t_0$;
(b) *uniformly bounded* if the value β in definition (a) does not depend on $t_0 \in \mathbb{R}_+$.

Theorem 3.6. *Assume that for the polynomial system (3.28), there exists a continuous functional $V(t, \varphi)$ defined for $t \in \mathbb{R}_+$ and $\varphi \in S^*$ so that:*

(1) *there are functions $a(r)$, $b_1(r)$, and $b_2(r)$ of the K-class such that*

$$a(|\varphi(0)|) \leq V(t, \varphi) \leq b_1(|\varphi(0)|) + b_2(\|\varphi\|),$$

and for $r > H$, $a(r) - b_2(r) \to \infty$ as $r \to \infty$;
(2) *the conditions of Lemma 3.2 are satisfied and there exists $\alpha > 0$ such that for $\|\varphi\| < \alpha$, the estimate*

$$M^*(t_0, t) = (m - 1) \int_{t_0}^t \sum_{i=2}^m (V^{i-1}(t_0, \varphi)\psi_i(\xi))$$

$$\times \exp\left(\int_{t_0}^\xi (m - 1)\psi_1(\tau)d\tau \right) d\xi < 1$$

holds uniformly in $t_0 \in \mathbb{R}_+$.
Then, the solutions of the system (3.28) are uniformly bounded.

Proof. For $\alpha > 0$, from condition (2) of Theorem 3.6, we choose $\beta(\alpha) > 0$ so that $b_1(\alpha) < a(\beta) - b_2(\beta)$. Let $x(t_0, \varphi)$ be a solution to the system (3.28) such that $\|\varphi\| < \alpha$ and $|x(t, t_0, \varphi)| = \beta$ for some $t > t_0$. Then, there are values t_1, t_2, and $t_0 \leq t_1 < t_2$ such that $|x(t_1, t_0, \varphi)| = \alpha$, $|x(t_2, t_0, \varphi)| = \beta$, and $\alpha < |x(t, t_0, \varphi)| < \beta$ for $t \in (t_1, t_2)$. In this case, we can assume that $|x(t, t_0, \varphi)| < \beta$ for $t \in [t_0, t_2)$ and hence $|x_t(t_0, \varphi)| < \beta$. From the bound (3.32), under condition (2) of Theorem 3.6, we have $V(t_2 \, x_{t_2}(t_0, \varphi)) \leq V(t_1, x_{t_1}(t_0, \varphi))$ for $t \in [t_1, t_2]$ and $x_t(t_0, \varphi) \in S^*$. The conditions of Theorem 3.6 imply the estimate $a(\beta) \leq b_1(\alpha) + b_2(\beta)$, which contradicts the choice of $\beta(\alpha) > 0$, so that $\beta_1(\alpha) < a(\beta) - b_2(\beta)$. Hence, $|x_t(t_0, \varphi)| < \beta(\alpha)$ uniformly in $t_0 \in \mathbb{R}_+$. This proves Theorem 3.6. $\square$

3.3.3 Method of Lyapunov–Razumikhin functions

If the functional $V(t, \varphi)$ does not depend on $\varphi \in C_H$, then it becomes a function, $V(t, x) : \mathbb{R}_+ \times \mathbb{R}^n \in \mathbb{R}_+ \to \mathbb{R}$, $V(t, 0) = 0$ for all $t \in \mathbb{R}_+$. The total derivative of the function $V(t, x)$ on the solutions of the system (3.28) is calculated using the formula

$$V'(t, h) = \frac{\partial V}{\partial t}(t, x) + \left(\frac{\partial V}{\partial x}(t, x)\right)^T f(t, h) \qquad (3.35)$$

for all $t \in \mathbb{R}_+$, $h = (x, x_t) \in B_r \times C_H$, where $B_r = \{x \in \mathbb{R}^n : |x| < r\}$.

Together with the derivative (3.35) of the function $V(t, x)$, the following set is considered (see [51, 56] and the bibliography therein):

$$\Omega_t(V) = \left\{ h \in C_H : \sup_{-\tau \leq s < 0} V(t + s, x(s)) \leq V(t, x) \right\}.$$

The application of the Lyapunov function method for a *polynomial system with aftereffect*, (3.28), is based on the following statement.

Lemma 3.3. *Let the Lyapunov function $V(t, x)$ be constructed for the system (3.28), for which $V'(t, h)$ on the solutions of the system (3.28) satisfies the estimate $V'(t, h) \leq 0$ for all $t_0 \in \mathbb{R}_+$ and $h \in \Omega_t(V)$. Then,*

$$V(t, x(t, \varphi)) \leq \sup_{-\tau \leq s < 0} V(t + s, x(t_0 + s, t_0, \varphi)) \qquad (3.36)$$

for all $t \in J \subseteq \mathbb{R}_+$.

Proof (see [56, p. 82]).

Consider the estimate of the total derivative (3.36) of the function $V(t, x)$ in the form

$$V'(t, h) \leq \sum_{i=1}^{m} \xi_i(t) V^i(t, h) \tag{3.37}$$

for all $t \in \mathbb{R}_+$ and $h \in \Omega_t(V)$. Here, the functions $\xi_i(t)$ have the same properties as in Lemma 3.2.

Lemma 3.4. *Suppose that for the system (3.28), the Lyapunov function $V(t, x)$ is constructed, for the total derivative, the estimate (3.37) holds. Then, the boundary of the variation of the function $V(t, x)$ on the solutions of the system (3.28) is estimated using the inequality*

$$V(t, x_t(t_0, \varphi)) \leq V(t_0, \varphi) \exp \left(\int_{t_0}^{t} \xi_1(s) ds \right) (1 - M^*(t_0, t))^{-\frac{1}{m-1}} \tag{3.38}$$

for all $t \geq t_0$, for which

$$M^*(t_0, t) = (m-1) \int_{t_0}^{t} \sum_{i=2}^{m} V^{i-1}(t_0, \varphi) \xi_i(s)$$

$$\times \exp \left(\int_{t_0}^{s} (m-1) \xi_1(\gamma) d\gamma \right) ds < 1.$$

The proof of Lemma 3.4 is similar to that of Lemma 3.1.

Corollary 3.4. *If $m = 2$ in Lemma 3.4, then (3.38) becomes*

$$V(t, x_t(t_0, \varphi)) \leq V(t_0, \varphi) \exp \left(\int_{t_0}^{t} \xi_1(s) ds \right) (1 - L^*(t_0, t))^{-1} \tag{3.39}$$

for all $t \geq t_0$, for which

$$L^*(t_0, t) = V(t_0, \varphi) \int_{t_0}^{t} \xi_2(s) \exp \left(\int_{t_0}^{s} \xi_1(\gamma) d\gamma \right) ds < 1.$$

3.3.3.1   Stability of solutions

Let us show that the following assertion holds.

Theorem 3.7. *Let a differentiable Lyapunov function $V(t, x)$ be constructed for the system (3.28) for which:*

(1) *all conditions of Lemma 3.4 are satisfied for all $t \geq t_0$;*
(2) *there exists a function $a \in K$-class such that $a(|x|) \leq V(t, x)$ for all $t \geq t_0$ and $x \in B_r$;*

(3) *there exists a constant $0 < k < \infty$ such that*

$$\exp\left(\int_{t_0}^{t} \bar{\xi}_1(s)ds\right)(1 - M^*(t_0, t))^{-\frac{1}{m-1}} \leq k$$

for all $t \geq t_0$.

Then, the zero solution of the polynomial system (3.28) is stable.

Proof. Let $t_0 \geq 0$ and $\varepsilon \in (0, r)$ be given. It follows from the condition $V(t, 0) = 0$ and the continuity of the function $V(t, x)$ that there exists $\delta = \delta(t_0, \varepsilon) > 0$ such that

$$\max\left(\sup_{-\tau \leq s < 0} V(t_0 + s, x(s)), \; kV(t_0, \varphi)\right) < a(\varepsilon)$$

for all $t \geq t_0$, $(x, \varphi) \in B_\delta$. Since under conditions (1) and (3) of Theorem 3.7, the inequality

$$V(\dot{\cdot}, x_t(\varphi)) - kV(t_0, \varphi) \leq 0 \tag{3.40}$$

holds for all $t \geq t_0$, the conditions of Lemma 3.4 are satisfied and for $\|\varphi\| < \delta$, we have the estimate $V(t, x_t(\varphi)) < a(\varepsilon)$. Hence, according to condition (2) of Theorem 3.7, we find that $|x(t, t_0, \varphi)| < \varepsilon$ for all $t \geq t_0$. This proves Theorem 3.7. □

Theorem 3.8. *Let the Lyapunov function $V(t, x)$ be constructed for the system (3.28) for which:*

(1) *all conditions of Lemma 3.4 are satisfied for all $t \geq t_0$;*
(2) *there exist comparison functions $a, b \in K$-class such that*

$$a(|x|) \leq V(t, x) \leq b(|x|)$$

for all $t \geq t_0$ and $x \in B_r$;
(3) *the inequality (3.38) holds uniformly in $t_0 \in \mathbb{R}_+$ with the function*

$$\tilde{M}(t_0, t) = (m - 1)\int_{t_0}^{t} \sum_{i=2}^{m} (b^i(\delta)\xi_i(s))$$

$$\times \exp\left(\int_{t_0}^{s} (m-1)\xi_1(\gamma)d\gamma\right)ds < 1;$$

(4) *condition (3) with the function $\tilde{M}(t, t_0)$ from Theorem 3.8 is satisfied uniformly in $t_0 \in \mathbb{R}_+$.*

Then, the zero solution of the polynomial system (3.28) is uniformly stable.

Proof.   The proof of Theorem 3.7 is similar to that of Theorem 3.8.   □

3.3.4   Applications

In this section, we consider the application of the method of Lyapunov–Razumikhin functions to the problem of the stability of a polynomial system of a special form and the stability on a finite interval of a *cubic polynomial equation*.

Example 3.3. Consider a polynomial equation with a special aftereffect:

$$\frac{dx}{dt} = f_0(t, x) + f_1(t, x_t)x + f_2(t, x_t)x^2 + f_3(t, x_t)x^3, \qquad (3.41)$$

where

$$f_0(t, x) = \bar{f}_0(t)x;$$

$$f_1(t, x_t) = x^3 \left(\sum_{i=1}^{k} \bar{f}_1^i(t)x(s_i) \right);$$

$$f_2(t, x_t) = x^4 \left(\sum_{i=1}^{k} \bar{f}_2^i(t)x^2(s_i) \right);$$

$$f_3(t, x_t) = x^5 \left(\sum_{i=1}^{k} \bar{f}_3^i(t)x^3(s_i) \right).$$

Here, $\bar{f}_1^i$, $\bar{f}_2^i$, and $\bar{f}_3^i$ are continuous functions on any finite interval and $\bar{f}_0(t) < 0$ for all $t \in \mathbb{R}_+$, $0 < \tau_1 < \tau_2 < \cdots < \tau_k$, $x(t) \in \mathbb{R}^1$, $x(s_i) \in C_H$, and $-\tau_i \leq s_i < 0$. With the equation (3.41), we consider the Lyapunov function $V(x) = \frac{1}{2}x^2$ and the set

$$\Omega(V) = \{(x, x(\cdot)) \in C_H : |x(s)| \leq |x| \quad \text{when } -\tau \leq s < 0\}.$$

Assume that there are integrable functions $\bar{f}_0(t), \ldots, \bar{f}_3(t)$ such that

$$\left| \sum_{i=1}^{k} \bar{f}_1^i(t)x(s_i) \right| \leq \bar{f}_1(t), \quad -\tau_i \leq s_i < 0;$$

$$\left| \sum_{i=1}^{k} \bar{f}_2^i(t)x^2(s_i) \right| \leq \bar{f}_2(t), \quad -\tau_i \leq s_i < 0; \qquad (3.42)$$

$$\left| \sum_{i=1}^{k} \bar{f}_3^i(t)x^3(s_i) \right| \leq \bar{f}_3(t), \quad -\tau_i \leq s_i < 0.$$

It is easy to see that under the conditions (3.42),

$$V'(x_t(t_0, \varphi)) \leq 2f_0(t)V(x_t(t_0, \varphi)) + 4\bar{f}_1(t)V^2(x_t(t_0, \varphi))$$

$$+ 8\bar{f}_2(t)V^3(x_t(t_0, \varphi)) + 16\bar{f}_3(t)V^4(x_t(t_0, \varphi)); \qquad (3.43)$$

$$V(\varphi(\theta)) \leq V(\varphi(0)), \ \theta \in [-\tau, 0],$$

where $\tau = \max_i \tau_i$, $i = 1, 2, \ldots, k$. Applying Lemma 3.4 to the inequality (3.43), we obtain an estimate for the function $V(x)$ on solutions of the equation (3.41) with the initial functions $\varphi(\theta) \in \Omega(V)$, $\theta \in [-\tau, 0]$ as

$$V(x_t(t_0, \varphi)) \leq V(\varphi) \exp\left(\int_{t_0}^t \bar{f}_0(s)ds\right)(1 - \bar{M}(t_0, t))^{-\frac{1}{3}} \qquad (3.44)$$

for all $t \geq t_0$, for which

$$\bar{M}(t_0, t) = 3\int_{t_0}^t \sum_{i=2}^4 (V^i(\varphi)\bar{f}_i(s)) \exp\left(3\int_{t_0}^s \bar{f}_0(\gamma)d\gamma\right)ds < 1. \qquad (3.45)$$

Condition (2) from Theorem 3.8 for the function $V(x)$ has the form $r^2 \leq V(x) \leq \beta r^2$ for any $\beta > 0$ and $r^2 = |x|^2$. It follows from the estimate (3.44) that the zero solution of the polynomial equation (3.41) will be uniformly stable if for any $\varepsilon > 0$, there exists $\delta = \delta(\varepsilon) > 0$ such that

$$\exp\left(\int_{t_0}^t \bar{f}_0(s)ds\right)(1 - \bar{M}(t_0, t))^{-\frac{1}{3}} < \frac{\varepsilon}{\delta}$$

for all $t \geq t_0$ and

$$\bar{M}(t_0, t) = 3\int_{t_0}^t \sum_{i=2}^4 \left(\delta^i(\varepsilon)\bar{f}_i(s)\right) \exp\left(3\int_{t_0}^s \bar{f}_0(\gamma)d\gamma\right)ds < 1$$

uniformly in $t_0 \in \mathbb{R}_+$.

Example 3.4. Consider a *cubic polynomial equation* with aftereffect

$$\frac{dx}{dt} = f_0(t, x_t)x + f_1(t, x_t)x^3, \qquad (3.46)$$

where $x \in \mathbb{R}$, $f_0 : \mathbb{R}_+ \times C_H \to \mathbb{R}$, $f_1 : \mathbb{R}_+ \times C_H \to \mathbb{R}_+$, and $f_0(t, 0) = f_1(t, 0) = 0$ for all $t \in \mathbb{R}_+$. We use the Lyapunov function $V(x) = \frac{1}{2}x^2$. It is easy to see that

$$V'(x_t(t_0, \varphi)) = f_0(t, x_t)x^2 + f_1(t, x_t)x^4$$

$$= 2f_0(t, x_t)V(x) + 4f_1(t, x_t)V^2(x). \qquad (3.47)$$

Let there exist functions $\bar{\psi}_i(t)$, $i = 1, 2$, such that

$$
\begin{aligned}
\|f_0(t, x_t)\|_{C_H} &\leq \bar{\psi}_1(t), \\
\|f_1(t, x_t)\|_{C_H} &\leq \bar{\psi}_2(t)
\end{aligned}
\tag{3.48}
$$

for all $t \geq t_0$ and $x_t \in C_H$, $\|x_t\| < H$. Here, the functions $\bar{\psi}_i(t)$, $i = 1, 2$, have the same properties as in Corollary 3.5. In view of the estimates (3.48), the relation (3.47) takes the form of the inequality

$$
\begin{aligned}
V'(x_t(t_0, \varphi)) &\leq 2\bar{\psi}_1(t)V(x_{t_0}(t, \varphi)) + 4\bar{\psi}_2(t)V^2(x_{t_0}(t, \varphi)), \\
V(\varphi(\theta)) &\leq V(\varphi(0)), \quad \theta \in [-\tau, 0],
\end{aligned}
\tag{3.49}
$$

for all $t \geq t_0$ and $\varphi(\theta) \in C_H$.

Applying Corollary 3.5 to the inequality (3.49), we obtain the estimate

$$
V(t, x_t(t_0, \varphi)) \leq V(t_0, \varphi) \exp\left(2 \int_{t_0}^{t} \bar{\psi}_1(s)ds \right) (1 - \tilde{L}(t_0, t))^{-1}
\tag{3.50}
$$

for all $t \geq t_0$, for which

$$
\tilde{L}(t_0, t) = V(t_0, \varphi) \int_{t_0}^{t} \bar{\psi}_2(s) \exp\left(2 \int_{t_0}^{s} \bar{\psi}_1(\gamma)d\gamma \right) ds < 1.
\tag{3.51}
$$

Definition 3.9. The solutions $x_t(t_0, \varphi)$ of the system (3.46) are (c_1, c_2, t_0, T)-stable if for given values $0 < c_1 < c_2$, $t_0 \in \mathbb{R}_+$, and T, under the condition $V(t_0, \varphi) < c_1$, the estimate $V(t, x_t(t_0, \varphi)) < c_2$ holds for all $t \in [t_0, t_0 + T]$.

It follows from the estimate (3.50) and the condition (3.51) that the solution $x_t(t_0, \varphi)$ of the system (3.46) is (c_1, c_2, t_0, T)-stable if

$$
\left(2 \int_{t_0}^{t} \bar{\psi}_1(s)ds \right) (1 - \hat{L}(t_0, t))^{-1} < \frac{c_2}{c_1},
$$

and

$$
\hat{L}(t_0, t) = \int_{t_0}^{t} \bar{\psi}_2(s) \exp\left(2 \int_{t_0}^{s} \psi_1(\gamma)d\gamma \right) ds < \frac{1}{4}
$$

for all $t \in [t_0, t_0 + T]$.

3.4 Stabilisation of the Motion of a Polynomial System

In this section, a new approach to solving the problem of stabilising the equilibrium state of non-autonomous polynomial systems is presented.

3.4.1 Statement of the problem

We consider a controlled polynomial system

$$\frac{dx}{dt} = f(t, x, u), \tag{3.52}$$

$$x(t_0) = x_0, \tag{3.53}$$

where $x(t) \in \mathbb{R}^n$ is the system state vector and the control $u = u(t, x) \in U \subseteq \mathbb{R}^m$.

Here, U is a given set of controls for all $(t, x) \in \mathbb{R}_+ \times D$, where $D \subseteq \mathbb{R}^n$ is an open domain in $\mathbb{R}^n$ containing the point $x = 0$.

The *control* $u(t, x)$ is assumed to be *admissible*, that is:

(1) the components of the vector function $u(t, x)$ are defined and piecewise continuous on any finite interval $[t_0, t_1] \subset \mathbb{R}_+$ for all $x \in D\backslash\{0\}$;
(2) $u(t, x) \in U$ for all $t \in [t_0, t_1]$;
(3) $u(t, x) = 0$ for all $t \in [t_0, t_1]$ if and only if $x = 0$;
(4) $u(t, x) < \infty$ for all $(t, x) \in \mathbb{R}_+ \times D\backslash\{0\}$.

In addition, we assume that for a given admissible control $u(t, x)$ and the initial conditions (3.53), the polynomial system of equations (3.52) has a unique continuous solution $x(t) = x(t, t_0, x_0)$ for all $t \in \mathbb{R}_+$. Such a *solution* is called *admissible* (see [146, p. 14]).

The vector function $f \in C(\mathbb{R}_+ \times \mathbb{R}^n \times U, \ \mathbb{R}^n)$ has homogeneous polynomials in n unknowns with reduced polynomials with respect to $x_1, x_2, \ldots, x_n$ and $f(t, 0, 0) = 0$ for all $t \in \mathbb{R}_+$.

Of interest is the problem of constructing a control $u = \bar{u}(t, x)$ such that the equilibrium state of the system of equations (3.52) stabilises to a stable or asymptotically stable one.

3.4.2 Preliminary results

Assume that in the system (3.52), the right-hand side is linear in control and has the form $f(t, x, u) = f_0(t, x) + f_1(t, x)u(t, x)$, where $f_0 \in C(\mathbb{R}_+ \times \mathbb{R}^n, \ \mathbb{R}^n)$ and $f_1 \in C(\mathbb{R}_+ \times \mathbb{R}^n, \ \mathbb{R}^n)$, $u \in \bar{U}$, $\bar{U} = (u(t, x) : \mathbb{R}_+ \times \mathbb{R}^n \to \mathbb{R})$.

Consider the *polynomial system* of equations of perturbed motion

$$\frac{dx}{dt} = f_0(t, x) + f_1(t, x)u(t, x), \tag{3.54}$$

$$x(t_0) = x_0. \tag{3.55}$$

For a *nominal polynomial system* of equations

$$\frac{dx}{dt} = f_0(t,x), \quad f_0(t,0) = 0, \tag{3.56}$$

construct a differentiable Lyapunov function $V \in C(\mathbb{R}_+ \times \mathbb{R}^n, \mathbb{R}_+)$, $V(t,x) > 0$, for all $x \in D\backslash\{0\}$, and assume that $\dfrac{\partial V(t,x)}{\partial t} = 0$ for $x = 0$. Let us introduce the notation

$$w(t,x) = \frac{\partial V}{\partial t}(t,x) + \left(\frac{\partial V}{\partial x}(t,x)\right)^T f_0(t,x);$$

$$p_1(t,x) = \left(\frac{\partial V}{\partial x}(t,x)\right)^T f_1(t,x).$$

Together with the function $V(t,x)$, we will consider its total derivative along the solutions of the system (3.52) under the initial conditions (3.53),

$$\frac{dV}{dt}(t,x) = \frac{\partial V}{\partial t}(t,x) + \left(\frac{\partial V}{\partial x}(t,x)\right)^T (f_0(t,x) + f_1(t,x)u(t,x)), \tag{3.57}$$

where

$$\frac{dV}{dt}(t,x) : \mathbb{R}_+ \times \mathbb{R}^n \to \mathbb{R}.$$

Let us show that the following assertion holds.

Theorem 3.9. *For the polynomial system* (3.54), *let the following conditions be satisfied:*

(1) *for the system* (3.56), *a differentiable Lyapunov function $V(t,x)$ with continuous and bounded partial derivatives is constructed;*
(2) *for all $(t,x) \in \mathbb{R}_+ \times D\backslash\{0\}$, the condition $p_1(t,x) \neq 0$;*
(3) *in the range $(t,x) \in \mathbb{R}_+ \times D\backslash\{0\}$, the following condition is fulfilled:*

$$\inf_{u \in U} \left[\frac{\partial V}{\partial t}(t,x) + \left(\frac{\partial V}{\partial x}(t,x)\right)^T f(t,x,u)\right] < 0.$$

Then, there is an admissible control $u = \bar{u}(t,x) \in \bar{U}$ for $(t,x) \in \mathbb{R}_+ \times D\backslash\{0\}$, which stabilises the motion of the system (3.54) *to an asymptotically stable one.*

Proof. Under conditions (1) and (2) of Theorem 3.9, we consider the Lyapunov function $V \in C(\mathbb{R}_+ \times \mathbb{R}^n, \mathbb{R}_+)$ and its total derivative due to the system (3.54). Taking into account some results of the papers [8, 162], we construct the control $\bar{u}(t, x)$ in the form

$$\bar{u}(t, x) = -\frac{w(t, x) + \sqrt{w^2(t, x) + p_1^2(t, x)}}{p_1(t, x))}. \tag{3.58}$$

It is easy to check that under this control, the total derivative with respect to t of the Lyapunov function $V(t, x)$ along the solutions of the system (3.54) satisfies the estimate

$$\frac{dV}{dt}(t, x) \le -\sqrt{w^2(t, x) + p_1^2(t, x)} < 0 \tag{3.59}$$

for all $(t, x) \in \mathbb{R}_+ \times D \backslash \{0\}$.

The inequality (3.59) together with condition (3) of Theorem 3.9 is sufficient for applying Lyapunov's theorem [76] on the asymptotic stability of the state $x = 0$ of the system (3.54).

It is easy to see from the formula (3.58) for the control $\bar{u}(t, x)$ that it satisfies conditions (1)–(4) given above for *admissible controls*.   □

3.4.3   Conditions for stabilising the motion of a polynomial system

In this section, based on the method of Lyapunov functions (see [76, 169]), we find a control and establish sufficient conditions under which the motion of a polynomial system stabilises to various types of stability or equi-boundedness.

Assume that the right-hand side of the system of equations (3.52) has the form

$$f(t, x, u) = f_0(t, x) + f_1(t, x)u(t, x) + \cdots + f_m(t, x)u^m(t, x), \tag{3.60}$$

and consider the system of polynomial equations of perturbed motion

$$\frac{dx}{dt} = f_0(t, x) + f_1(t, x)u(t, x) + \cdots + f_m(t, x)u^m(t, x), \tag{3.61}$$

where $f_i(t, x) \in C(\mathbb{R}_+ \times \mathbb{R}^n, \mathbb{R}^n), i = 0, 1, 2, \ldots, m$, are polynomial functions with respect to the variables $x_1, \ldots, x_n$, $(u, \ldots, u^m) \in \bar{U}$ and $f_i(t, 0) = 0$ for all $t \ge t_0$.

For a *nominal polynomial system* of equations

$$\frac{dx}{dt} = f_0(t, x), \tag{3.62}$$

$$x(t_0) = x_0, \tag{3.63}$$

where $x \in \mathbb{R}^n$ and $f_0 \in C(\mathbb{R}_+ \times \mathbb{R}^n, \mathbb{R}^n)$, we construct a differentiable Lyapunov function $V(t, x) > 0$ for all $(t, x) \in \mathbb{R}_+ \times D\backslash\{0\}$ and $V(t, 0) = 0$. As before, it is assumed that $\dfrac{\partial V(t, x)}{\partial t} = 0$ for $x = 0$.

Denote

$$w(t, x) = \frac{\partial V}{\partial t}(t, x) + \left(\frac{\partial V}{\partial x}(t, x)\right)^T f_0(t, x) : \mathbb{R}_+ \times \mathbb{R}^n \to \mathbb{R},$$

$$p_i(t, x) = \left(\frac{\partial V}{\partial x}(t, x)\right)^T f_i(t, x) : \mathbb{R}_+ \times \mathbb{R}^n \to \mathbb{R}, \quad i = 1, 2, \ldots, m.$$

Further, for $p_1(t, x) \neq 0$, for all $(t, x) \in \mathbb{R}_+ \times D\backslash\{0\}$, the control

$$\tilde{u}(t, x) = -\frac{w(t, x)}{p_1(t, x)} : \mathbb{R}_+ \times \mathbb{R}^n \to \mathbb{R}, \tag{3.64}$$

constructed for the truncated system of equations

$$\frac{dx}{dt} = f_0(t, x) + f_1(t, x)u(t, x), \tag{3.65}$$

will be considered as approximate for the system of equations (3.61).

3.4.4 Stabilisation of motion to steady state

Note that the control (3.64) stabilises the zero solution of the system (3.65) to a stable one.

Let us show that the following assertion holds.

Theorem 3.10. *Assume that for the nominal system* (3.62), *we construct a differentiable Lyapunov function $V(t, x)$ defined for all $0 \leq t < \infty$ and $x \in B_r\backslash\{0\}$, where $B_r = \{x \in \mathbb{R}^n : \|x\| < r\}$, on the basis of which the control* (3.64) *is constructed for the shortened system* (3.65). *If the following conditions are fulfilled:*

(1) *$V(t, x) = 0$ for all $t \geq t_0$ and for $x = 0$;*
(2) *there exists a function $a \in K$-class such that $a(\|x\|) \leq V(t, x)$ for all $(t, x) \in \mathbb{R}_+ \times B_r$;*
(3) *$\sum_{i=2}^{m} p_i(t, x)\tilde{u}^i(t, x) \leq 0$ for all $t \geq t_0$ and $x \in B_r\backslash\{0\}$, then the approximate control* (3.64) *stabilises the zero solution of the system* (3.61) *to a stable one.*

Proof. It follows from the conditions of Theorem 3.10 that under conditions (1)–(3), the total derivative with respect to t of the Lyapunov function $V(t,x)$ is less than or equal to zero, i.e., all the conditions of Lyapunov's theorem [76] on the stability of motion are satisfied. $\qquad\square$

Theorem 3.11. *If, in the conditions of Theorem 3.10, condition (2) is replaced by*

$$(2)' \quad a(\|x\|) \le V(t,x) \le b(\|x\|),$$

where the functions $a, b \in K$-class, then the approximate control (3.64) stabilises the zero solution of the system (3.61) to a uniformly stable one.

Proof. Under the conditions of Theorem 3.10, with a Lyapunov function satisfying estimates (2)′, all the conditions of Theorem 3.11 from the monograph [169] are satisfied (see also Corollary 2 in the monograph [177, p. 57]); therefore, the approximate control (3.64) stabilises the zero solution of the system (3.61) to a uniformly stable one. Next, we consider the problem of stabilisation of the motion of the system (3.61) by the control (3.64) to an asymptotically stable one. Let us show that the following assertion is true. $\qquad\square$

Theorem 3.12. *If, in the conditions of Theorem 3.10, condition (2) is replaced by condition (2)′ from Theorem 3.11 and condition (3) is replaced by the following:*
(3)′ there exist $1 < k < m$ and a function $c \in K$-class such that

$$\sum_{i=2}^{k} p_i(t,x) \tilde{u}^i(t,x) \le -c(\|x\|)$$

for all $t \ge t_0$ and $x \in B_r \backslash \{0\}$, then the approximate control (3.64) stabilises the zero solution of the system

$$\frac{dx}{dt} = f_0(t,x) + f_1(t,x)u(t,x) + \cdots + f_k(t,x)u^m(t,x) \qquad (3.66)$$

to a uniformly asymptotically stable one.

Proof. Under the conditions of Theorem 3.12, all the conditions of Theorem 3.11 from the monograph [169] are satisfied; therefore, the approximate control (3.64) stabilises the zero solution of the system to a uniformly asymptotically stable one. $\qquad\square$

Corollary 3.5. *If the conditions of Theorem* 3.12 *are satisfied with condition* (3)′ *replaced by the following:*
(3)″ *there exist* $1 < k < m$ *and a constant* $q > 0$ *such that*

$$\sum_{i=2}^{k} p_i(t,x)\tilde{u}^i(t,x) \leq -qV(t,x)$$

for all $t \in \mathbb{R}_+$ *and* $x \in B_r\backslash\{0\}$, *then the approximate control* (3.64) *stabilises the zero solution of the system* (3.66) *to a uniformly asymptotically stable one.*

Next, we rewrite the polynomial system (3.66) as

$$\frac{dx}{dt} = \sum_{i=0}^{k} f_i(t,x)\tilde{u}^i(t,x) + \sum_{i=k+1}^{m} f_i(t,x)\tilde{u}^i(t,x). \qquad (3.67)$$

Assume that for the values x in $\|x\| < \varepsilon$, there exists $\triangle(\varepsilon) > 0$ such that

$$\left\| \sum_{i=k+1}^{m} f_i(t,x)\tilde{u}^i(t,x) \right\| < \triangle(\varepsilon) \qquad (3.68)$$

for all $t \in \mathbb{R}_+$ and $x \in B_\varepsilon\backslash\{0\}$.

Let us show that an admissible solution $x(t) = x(t,t_0,x_0)$ of the system (3.66) satisfies the estimate $\|x(t,t_0,x_0)\| < \varepsilon$ for $\|x_0\| < \triangle(\varepsilon)$ and all $t \geq t_0$ if the approximate control (3.64) is constructed using the Lyapunov function satisfying the Lipschitz condition

$$|V(t,x) - V(t,x')| \leq L\|x - x'\| \qquad (3.69)$$

for all $(t,x) \in \mathbb{R}_+ \times B_\varepsilon\backslash\{0\}$, $L > 0$ is the Lipschitz constant. Let us show that the following assertion holds.

Theorem 3.13. *Let the control* (3.64) *constructed on the basis of a differentiable function* $V(t,x)$, *satisfying the estimate* (3.69), *stabilise the zero solution of the system* (3.65) *to uniformly asymptotically stable, and in addition, the conditions of Corollary* 3.5 *and the condition* (3.68) *be satisfied. Then the zero solution of the system* (3.66) *is stabilised by the control* (3.64) *to a stable one.*

Proof. We consider a function $V(t,x)$ that satisfies the estimate (3.69) and a control (3.64) constructed on the basis of this function. In Corollary 3.5, we take $q = 1$ and, for any $\varepsilon > 0$, choose $\delta_1(\varepsilon) > 0$ so that the inequality $b(\delta_1) < a(\varepsilon)$ holds. Here, $a(r)$ and $b(r)$ are functions of the Hahn class from condition (2)′ of Theorem 3.11. Next, we choose $0 < \triangle(\varepsilon) < \delta_1$ so small that $a(\delta_1) - L\triangle(\varepsilon) > 0$, where L is the Lipschitz constant from the bound (3.69). Let the condition (3.68) hold and an admissible solution $x(t) = x(t, t_0, x_0)$ for $\|x_0\| < \triangle(\varepsilon)$ satisfy the condition $\|x(t, t_0, x_0)\| = \varepsilon$ at some time $t \in \mathbb{R}_+$. Then, there are values $t_1, t_2 \in \mathbb{R}_+$ such that $\|x(t_1, t_0, x_0)\| = \delta_1$ and $\|x(t_2, t_0, x_0)\| = \varepsilon$, and for all $t \in (t_1, t_2)$, the solution $x(t)$ satisfies the estimate $\|x(t, t_0, x_0)\| < \varepsilon$. At the same time, under the conditions of Theorem 3.13, for all $t \in [t_1, t_2]$, we have the estimate

$$\frac{dV}{dt}(t, x) \leq -V(t, x(t))$$

$$+ L\left\|\sum_{i=k+1}^{m} f_i(t, x)\tilde{u}^i(t, x)\right\| \leq -a(\delta_1) + L\triangle(\varepsilon) \leq 0 \quad (3.70)$$

in the region $\|x\| < \varepsilon$. $\square$

From the condition $(2)'$ of Theorem 3.11 and the estimate (3.70), we obtain the inequalities

$$a(\varepsilon) \leq V(t_2, x(t_2)) \leq V(t_1, x(t_1)) \leq b(\delta_1).$$

This contradicts the choice of δ_1. Therefore, $\|x(t, t_0, x_0)\| < \varepsilon$ is true for all $t \geq t_0$. Thus, the control (3.64) stabilises the zero solution of the system (3.66) under the condition (3.68).

3.4.5   Stabilisation of motion to an equi-bounded one

Suppose that the vector functions $f_i(t, x)$, $i = 0, 1, \ldots, m$, in the system (3.66) are continuous on $\mathbb{R}_+ \times \mathbb{R}^n$ and that the Lyapunov function $V(t, x) > 0$ is defined for $0 \leq t < \infty$ and $\|x\| \geq r$, where r can be arbitrarily large.

We consider the system (3.66) in the context of the following definition.

Definition 3.10. An admissible solution $x(t, t_0, x_0)$ of the system (3.66) is *stabilised* by a control of the form (3.64) to an *equi-bounded* one if, for a given $\alpha > 0$, there exist two numbers $\beta(\alpha) > 0$ and $\triangle(\alpha) > 0$ such that for $\|x_0\| < \alpha$, the estimate $\|x(t, t_0, x_0)\| < \beta(\alpha)$ holds for all $t \geq t_0$, where $x(t, t_0, x_0)$ is an admissible solution to the system (3.66) under the condition

$$\left\|\sum_{i=k+1}^{m} f_i(t, x)\tilde{u}^i(t, x)\right\| < \triangle(\alpha) \quad (3.71)$$

for all $t \geq t_0$ and $\alpha < \|x\| < \beta(\alpha)$.

The following assertion holds.

Theorem 3.14. *Let the control* (3.64) *be constructed on the basis of the Lyapunov function* $V(t, x)$ *for which the following conditions are satisfied:*

(1) *there exist a function* $a \in KR$*-class and a function* $b \in K$*-class such that*

$$a(\|x\|) \leq V(t, x) \leq b(\|x\|)$$

for all $t \geq t_0$ *and* $x \in \mathbb{R}^n$;

(2) $|V(t, x) - V(t, x')| \leq k(\alpha)\|x - x'\|$, *where* $k(\alpha) > 0$;

(3) *there exists* $1 < k < m$ *such that*

$$\left\| \sum_{i=2}^{k} p_i(t, x)\tilde{u}^i(t, x) \right\| \leq -c(\|x\|),$$

where $c(r) > 0$ *is a continuous function.*

(4) *given* $\alpha > 0$, *there exist* $\triangle(\alpha) > 0$ *and* $\beta(\alpha) > 0$ *such that the condition* (3.71) *holds in the range* $\alpha < \|x\| < \beta(\alpha)$.

Then, the control (3.64) *stabilises the motion of the system* (3.66) *to an equi-bounded one in the sense of Definition 3.10.*

Proof.   Let $\alpha > 0$ be given. We choose $\beta(\alpha) > 0$ so that $a(\alpha) > b(\beta)$. Under conditions (2) and (3) of Theorem 3.14, for the values $\alpha \leq \|x\| \leq \beta(\alpha)$, there are constants $k(\alpha) > 0$ and $\lambda(\alpha) > 0$ such that

$$\sum_{i=2}^{k} p_i(t, x)\tilde{u}^i(t, x) \leq -\lambda(\alpha)$$

and

$$\frac{dV}{dt}(t, x) \leq \sum_{i=2}^{k} p_i(t, x)\tilde{u}^i(t, x) + k(\alpha) \left\| \sum_{i=k+1}^{m} p_i(t, x)\tilde{u}^i(t, x) \right\| \qquad (3.72)$$

for all $0 \leq t < \infty$ and $\alpha \leq \|x\| < \beta(\alpha)$. $\square$

Under condition (4) of Theorem 3.14, we choose $\triangle(\alpha)$ so that $\triangle(\alpha) \leq \lambda(\alpha)/k(\alpha)$. At the same time, from the estimate (3.72), we obtain

$$\frac{dV}{dt}(t, x) \leq 0 \qquad (3.73)$$

in the range $0 \leq t < \infty$ and $\alpha \leq \|x\| \leq \beta(\alpha)$ as soon as (3.68) holds. From the bound (3.73), for $\|x_0\| < \alpha$, it follows that the admissible solution

$x(t)$ satisfies the inequality $\|x(t)\| < \beta(\alpha)$ for all $t \in \mathbb{R}_+$. This proves Theorem 3.14.

3.4.6   Stabilisation of a scalar polynomial equation

A *scalar equation* of perturbed motion with *polynomial control* is considered in the form

$$\frac{dx}{dt} = f_0(t,x) + f_1(t,x)u(t,x) + \cdots + f_m(t,x)u^m(t,x), \qquad (3.74)$$

$$x(t_0) = x_0, \qquad (3.75)$$

where $x \in \mathbb{R}$, functions $f_i(t,x) \in C(\mathbb{R}_+ \times \mathbb{R},\ \mathbb{R})$ and are locally Lipschitz, control $(u,\ldots,u^m) \in \bar{U}$, where $\bar{U}$ is an admissible set of scalar controls.

It is assumed that the functions $f_i(t,x)$ for $i = 0,1,2,\ldots,m$ are homogeneous polynomials.

For the nominal equation

$$\frac{dx}{dt} = f_0(t,x), \qquad (3.76)$$

where $f_0 \in C(\mathbb{R}_+ \times \mathbb{R},\ \mathbb{R})$, $f_0(t,0) = 0$, for all $t \in \mathbb{R}_+$, let a differentiable function $V(t,x) > 0$ be constructed for $x \in \mathbb{R}\backslash\{0\}$ and $t \in \mathbb{R}_+$.

The total derivative with respect to t of the function $V(t,x)$ due to the equation (3.74),

$$\frac{dV}{dt}(t,x) = \frac{\partial V}{\partial t} + \frac{\partial V}{\partial x}(t,x)f_0(t,x) + \sum_{i=1}^{m}\left(\frac{\partial V}{\partial x}(t,x)f_i(t,x)\right)u^i(t,x),$$

$$(3.77)$$

is represented in the form

$$\frac{dV}{dt}(t,x) = w^*(t,x) + \sum_{i=1}^{m} p_i(t,x)u^i(t,x), \qquad (3.78)$$

where

$$w^*(t,x) = \frac{\partial V}{\partial t} + \frac{\partial V}{\partial x}(t,x)f_0(t,x),$$

$$p_i(t,x) = \frac{\partial V}{\partial x}(t,x)f_i(t,x), \quad i = 1,2,\ldots,m.$$

The right-hand side of the relation (3.78) is a polynomial of degree m with respect to the desired control $u^i(t,x)$, $i = 1,2,\ldots,m$.

Consider an algebraic equation with real coefficients

$$p_m(t,x)u^m + p_{m-1}(t,x)u^{m-1} + \cdots + p_1(t,x)u + w^*(t,x) = 0 \qquad (3.79)$$

in the range $(t,x) \in \mathbb{R}_+ \times \bar{D}$, where $\bar{D}$ is an open interval on R.

It is known (see [167]) that the equation (3.79) with literal coefficients has no radical solution for $m > 4$.

Assume that a solution to the algebraic equation (3.79) exists for all $(t,x) \in \mathbb{R}_+ \times \bar{D}$.

If $v_1, \ldots, v_m$ are real roots of the equation (3.79), then they lie in the interval $(-N, N)$, where

$$N = 1 + \frac{\sup\limits_{(t\in\mathbb{R}_+, x\in\bar{D})} \max\limits_{i\in[1,m-1]} (|p_i(t,x)|, |w^*(t,x)|)}{\sup\limits_{(t\in\mathbb{R}_+, x\in\bar{D})} |p_m(t,x)|}. \qquad (3.80)$$

Along with this formula, for the boundary of the roots of the algebraic equation (3.79), on the basis of known results (see [117, p. 214]), more precise boundaries can be specified.

The real roots of the algebraic equation (3.79) form the set of the values of *admissible controls* $\bar{U}$. The admissible control $\bar{u}(t,x)$ is in the interval

$$-N \leq \bar{u}(t,x) \leq N$$

for all $t \in [t_1, t_2], t_1, t_2 \in \mathbb{R}_+$.

Remark 3.6. It is known that an admissible control in a real system has a certain physical meaning. For example, in the problem of the flight range of a rocket with limited acceleration, one of the control components is the rocket's mass consumption rate (see [74, pp. 78–84]).

If the condition $V(t,x) > 0$ is satisfied, together with the condition

$$w^*(t,x) + \sum_{i=1}^{m} p_i(t,x)\bar{u}^i(t,x) = 0, \qquad (3.81)$$

then the solution $x = 0$ of the equation (3.74) under the control $\bar{u}(t,x)$ will be stable due to Lyapunov's stability theorem (see [76]).

Remark 3.7. It is assumed that the roots $v_i, \ldots, v_m$ of the equation (3.79) satisfy the condition $v_i(t,x) \cap v_j(t,x) = \oslash$ for all $i \neq j$, $i, j = 1, 2, \ldots, m$, and for all $(t,x) \in \mathbb{R}_+ \times \bar{D}\backslash\{0\}$.

3.4.7   Motion stabilisation in a particular case

Let us continue the consideration of the equation (3.74) under some assumptions about the nominal system (3.76). Suppose that in the relation (3.78), the function $w^*(t, x) \leq 0$ in the range $(t, x) \in \mathbb{R}_+ \times \bar{D}$, i.e., the zero solution of the equation (3.27) is Lyapunov-stable.

Moreover, the relation (3.78) implies the estimate

$$\frac{dV}{dt}(t, x) \leq \sum_{i=1}^{m} p_i(t, x) u^i(t, x) \qquad (3.82)$$

in the range $(t, x) \in \mathbb{R}_+ \times \bar{D}$.

From the right-hand side of the inequality (3.82), we select the algebraic equation

$$p_m(t, x) u^{m-1}(t, x) + p_{m-1}(t, x) u^{m-2}(t, x) + \cdots + p_1(t, x) = 0 \qquad (3.83)$$

and suppose that $p_m(t, x) \neq 0$ for all $(t, x) \in \mathbb{R}_+ \times \bar{D}$. We represent the equation (3.83) as

$$u^{m-1}(t, x) + \bar{p}_{m-1}(t, x) u^{m-2}(t, x) + \cdots + \bar{p}_1(t, x) = 0, \qquad (3.84)$$

where

$$\bar{p}_{m-1}(t, x) = \frac{p_{m-1}(t, x)}{p_m(t, x)}, \dots, \quad \bar{p}_1(t, x) = \frac{p_1(t, x)}{p_m(t, x)}.$$

If $v_1, \dots, v_{m-1}$ are the roots of the equation (3.84), then the control $\tilde{u}(t, x)$ is in the interval

$$-N \leq \tilde{u}(t, x) \leq N. \qquad (3.85)$$

Obviously, if the control $\tilde{u}(t, x)$ together with the inequality $w^*(t, x) \leq 0$ bring the relation (3.77) to the estimate

$$\frac{dV}{dt}(t, x) \leq \sum_{i=1}^{m-1} p_i(t, x) \tilde{u}^{m-1}(t, x) \leq 0, \qquad (3.86)$$

then the control $\tilde{u}(t, x)$ stabilises the solution $x = 0$ of the equation (3.74) to a stable one in the case of Lyapunov stability of the nominal polynomial equation (3.76).

Example 3.5. Consider the scalar polynomial equation

$$\frac{dx}{dt} = a_0(t)x + a_1(t)x^3 u + a_2(t)x^5 u^2 + a_3(t)x^7 u^3, \qquad (3.87)$$

$$x(t_0) = x_0, \qquad (3.88)$$

where $a_i(t) \in C(\mathbb{R}_+, \mathbb{R})$, $i = 1, 2, 3, 4$, $u \in U$.

For the nominal equation

$$\frac{dx}{dt} = a_0(t)x, \tag{3.89}$$

we consider the Lyapunov function $V(x) = x^2$, $x \in \mathbb{R}\backslash 0$. The function $w^*(t, x)$ has the form $w^*(t, x) = 2a_0(t)x^2$.

The functions $p_i(t, x)$, $i = 1, 2, 3$, have the form

$$p_1(t, x) = 2a_1(t)x^4, \ p_2(t, x) = 2a_2(t)x^6, \ p_3(t, x) = 2a_3(t)x^8. \tag{3.90}$$

Given (3.90), the relation (3.77) becomes

$$\frac{dV}{dt}(x(t)) \le 2a_1(t)x^4 u + 2a_2(t)x^6 u^2 + 2a_3(t)x^8 u^3. \tag{3.91}$$

Hence, we have the algebraic equation

$$p_1(t, x)u + p_2(t, x)u^2 + p_3(t, x)u^3 = 0$$

with one zero root.

Consider the equation

$$p_3(t, x)u^2 + p_2(t, x)u + p_1(t, x) = 0, \tag{3.92}$$

and suppose that $p_3(t, x) \neq 0$ for all $(t, x) \in \mathbb{R}_+ \times \bar{D}$. We represent the equation (3.92) as

$$u^2(t, x) + p(t, x)u(t, x) + q(t, x) = 0, \tag{3.93}$$

where

$$p(t, x) = p_2(t, x)/p_3(t, x), \ q(t, x) = p_1(t, x)/p_3(t, x).$$

Since the solutions of the equation (3.93) are the functions

$$u_1(t, x) = \frac{1}{2}(-p(t, x) + \sqrt{D}),$$

$$u_2(t, x) = \frac{1}{2}(-p(t, x) - \sqrt{D}),$$

where $D = p^2(t, x) - 4q(t, x)$, the control $\tilde{u}(t, x) = u_1(t, x)$ (or $\tilde{u}(t, x) = u_2(t, x)$) together with the inequality (3.86) stabilise the solution $x = 0$ of the equation (3.74) to a stable one.

Example 3.6. Consider the truncated scalar equation (3.74) as

$$\frac{dx}{dt} = f_0(t, x) + f_1(t, x)u(t, x), \tag{3.94}$$

where $x \in \mathbb{R}$, $f_0, f_1 \in C(\mathbb{R}_+ \times \mathbb{R}, \ \mathbb{R})$, $u(t, x) \in U$.

Applying the function $V(x) = \frac{1}{2}x^2$, we obtain expressions for

$$w^*(t,x) = xf_0(t,x) \quad \text{and} \quad p_1(t,x) = xf_1(t,x).$$

Let $p_1(t,x) > 0$ for all $(t,x) \in \mathbb{R}_+ \times \mathbb{R}\backslash\{0\}$. We choose the control $\bar{u}(t,x)$ according to the formula

$$\bar{u}(t,x) = -\frac{w^*(t,x) + |x|\sqrt{f_0^2(t,x) + x^2 f_1^4(t,x)}}{p_1(t,x)}. \tag{3.95}$$

It is easy to see that under the control (3.95), on the solution $x(t)$ of the equation (3.94), we have the estimate

$$\frac{dV}{dt}(x(t)) = -|x|\sqrt{f_0^2(t,x) + x^2 f_1^4(t,x)} < 0. \tag{3.96}$$

It follows from the inequality (3.96) that the control (3.95) stabilises the solution $x = 0$ of the equation (3.94) to an asymptotically stable one.

3.5   Analysis of Motion with Interval Initial Conditions

In this section, we propose a constructive application of nonlinear integral inequalities in estimating motions and obtaining conditions for stabilising the motion of a nonlinear system with many control elements under *interval initial conditions*.

3.5.1   Statement of the problem

The following nonlinear system of equations of controlled motion is considered:

$$\frac{dy}{dt} = F(t,y,u), \tag{3.97}$$

where $y \in \mathbb{R}^n$, $F \in C(\mathbb{R}_+ \times \mathbb{R}^n \times \mathbb{R}^m, \mathbb{R}^n)$, and $u(t) \in U \subset \mathbb{R}^m$ is the motion control. Let, for a fixed control $u_p(t) \in U$, the interval initial data for the solution $y(t)$ be given:

$$y(t_0) = (\underline{y}_0, \overline{y}_0) \in \mathbb{R}^n. \tag{3.98}$$

Under broad assumptions about the system (3.97), its solution $y = y_p(t) = y(t, t_0, y_0, u_p)$ exists on any finite interval $I \subset \mathbb{R}_+$.

Of interest is the problem of estimating motion under the interval initial conditions (3.98) and conditions for stabilising motion to an exponentially stable one.

In the system (3.97), we change the variables x and u according to the formulas $y = y_p + x$ and $u = u_p + v$. Then, the system (3.97) becomes

$$\frac{dx}{dt} = f(t, x, v), \tag{3.99}$$

where $f(t, x, v) = F(t, y_p + x, u_p + v) - F(t, y_p, u_p)$. In this case, $f(t, 0, 0) \equiv 0$, and for $v = 0$, the system (3.99) has the motion $x = 0$.

Further, the system of equations (3.99) will be considered in the form

$$\frac{dx}{dt} = f(t, x) + g(t, x, v), \tag{3.100}$$

where $f(t, x) = f(t, x, 0)$ and $g(t, x, v) = f(t, x, v) - f(t, x)$. Consider the problem of estimating the norm of a solution $x(t)$ under the interval initial conditions

$$x(t_0) = x_0 \in [\underline{x}_0, \overline{x}_0], \tag{3.101}$$

where $\underline{x}_0, \overline{x}_0$ are predefined values.

3.5.2 Estimates of motions of the system (3.100)

Here and in the following, the interval norm is applied (see Appendix A). We make the following assumptions about the system (3.100):

C_1. There exists an integrable non-negative function $m(t)$ for all $t \geq 0$ such that

$$\|f(t, x)\| \leq m(t)\|x\| \tag{3.102}$$

for all $(t, x) \in \mathbb{R}_+ \times \mathbb{R}^n$.

C_2. There exists an integrable non-negative function $l(t)$ and a constant $p > 1$ such that

$$\|g(t, x, v)\| \leq l(t)\|x\|^p \tag{3.103}$$

for all $(t, x, v) \in \mathbb{R}_+ \times \mathbb{R}^n \times U$. The following statement is valid.

Lemma 3.5. *If the conditions of assumptions C_1 and C_2 are satisfied, then the norm of the solution $x(t)$ of the equation (3.100) satisfies the estimate*

$$\|x(t)\| \leq \frac{\|x_0\|_2 \exp\left(\int_0^t m(s)\,ds\right)}{\left(1 - (p-1)\|x_0\|_2^{p-1}\int_0^t l(s)\exp\left[(p-1)\int_0^s m(\tau)\,d\tau\right]ds\right)^{\frac{1}{p-1}}}$$

for all $t \geq s \geq 0$, for which

$$(p-1)\|x_0\|_2^{p-1} \int_0^t l(s) \exp\left[(p-1)\int_0^s m(\tau)\,d\tau\right] ds < 1.$$

Proof.   We represent the equation (3.100) as

$$x(t) = x(t_0) + \int_{t_0}^t f(s, x(s))\,ds + \int_{t_0}^t g(s, x(s), v(s))\,ds \qquad (3.104)$$

for all $t \geq t_0 \geq 0$ and $x(t_0) = x_0 \in [\underline{x}_0, \overline{x}_0]$. From the relation (3.104), under assumptions C_1 and C_2, we obtain the estimate

$$\|x(t)\| \leq \|x_0\|_2 + \int_0^t m(s)\|x(s)\|\,ds + \int_0^t l(s)\|x(s)\|^p\,ds. \qquad (3.105)$$

Let $\|x(t)\| = \psi(t)$ for all $t \geq 0$. From the inequality (3.104), it follows that

$$\psi(t) \leq \|x_0\|_2 + \int_0^t (m(s) + l(s)\psi^{p-1}(s))\psi(s)\,ds. \qquad (3.106)$$

$\square$

Further, the proof of Lemma 3.5 is similar to the proof of Lemma 1.6, and for this reason, it is omitted here.

Lemma 3.6. *Let there exist functions $m_k(t)$, $k = 1, 2, \ldots, n$, integrable for all $t \geq 0$ such that*

$$\|f(t, x) + g(t, x, v)\| \leq \sum_{k=1}^n m_k(t)\|x\|^k \qquad (3.107)$$

for all $(t, x, v) \in \mathbb{R}_+ \times \mathbb{R}^n \times V$. Then, the norm of the solution $x(t)$ of the system (3.100) satisfies the estimate

$$\|x(t)\|$$

$$\leq \frac{\|x_0\|_2 \exp\left(\int_0^t m_1(s)\,ds\right)}{\left(1 - (n-1)\int_0^t \sum_{k=2}^n \|x_0\|_2^{k-1} m_k(s) \exp\left(\int_0^s (n-1)m_1(\tau)\,d\tau\right) ds\right)^{\frac{1}{n-1}}}$$

$$(3.108)$$

for all $t \geq s \geq 0$, for which

$$(n-1)\int_0^t \sum_{k=2}^n \|x_0\|_2^{k-1} m_k(s) \exp\left(\int_0^s (n-1)m_1(\tau)\,d\tau\right) ds < 1.$$

Proof.   From the relation (3.104), under the condition (3.107), we obtain the estimate

$$\|x(t)\| \le \|x_0\|_2 + \int_0^t \sum_{k=1}^n m_k(s)\|x^k(s)\|\,ds, \tag{3.109}$$

which is represented in the pseudo-linear form

$$\|x(t)\| \le \|x_0\|_2 + \int_0^t \left(m_1(s) + \sum_{k=2}^n m_k(s)\|x^{k-1}(s)\| \right) \|x(s)\|\,ds \tag{3.110}$$

for all $t \ge 0$. Applying the Gronwall–Bellman lemma to the inequality (3.110), we obtain the estimate

$$\|x(t)\| \le \|x_0\|_2 \exp\left(\int_0^t \left(m_1(s) + \sum_{k=2}^n m_k(s)\|x^{k-1}(s)\| \right) ds \right) \tag{3.111}$$

for all $t \ge 0$.   □

Applying the estimation technique from the proof of Lemma 2.7 to the inequality (3.111), we obtain the required estimate (3.108).

3.5.3   Motion stabilisation

Consider a *nonlinear system* with *many control* elements:

$$\frac{dx}{dt} = A(t)x(t) + \sum_{i=1}^m G_i(t, x(t))u_i(t) + Bu_0(t),$$

$$y(t) = Cx(t), \tag{3.112}$$

$$x(t_0) = x_0 \in [\underline{x}_0, \overline{x}_0],$$

where $x \in \mathbb{R}^n$, $A(t)$ is an $n \times n$ matrix with continuous elements on any finite interval of motion, $G_i(t, x)$ is an $n \times m$ matrix, the control vectors $u_i(t) \in \mathbb{R}^m$ for all $i = 1, 2, \ldots, m$, B is an $n \times m$ matrix and the control $u_0(t) \in \mathbb{R}^m$, C is a constant $n \times n$ matrix, and x_0 is an interval vector of the initial state of the system (3.112). We make the following assumptions about the system (3.112):

C_3. The functions $G_i(t, 0) = 0$, $i = 1, 2, \ldots, m$, for all $t \ge 0$.
C_4. There exists a constant $n \times m$ matrix K_0 such that for the system

$$\frac{dy}{dt} = (A(t) - BK_0C)y,$$

the fundamental matrix $\Phi(t)$ satisfies the estimate

$$\|\Phi(t)\Phi^{-1}(s)\| \leq Me^{-\alpha(t-s)} \tag{3.113}$$

for $t \geq s \geq t_0$, where M and α are some positive constants.

C_5. There are constants $\gamma_i > 0$ and $q > 1$ such that

$$\|G_i(t, x)\| \leq \gamma_i \|x\|^q$$

for all $i = 1, 2, \ldots, m$.

The following assertion holds.

Theorem 3.15. *Let assumptions C_3–C_5 be satisfied. Then, the controls*

$$u_i(t) = -K_i y(t), \quad i = 1, 2, \ldots, m, \quad u_0(t) = -K_0 y(t) \tag{3.114}$$

stabilise the motion of the system (3.112) to an exponentially stable one.

Proof.   Let $m < n$ and the controls $u_i(t) = -K_i y(t)$ and $u_0(t) = -K_0 y(t)$ be applied to stabilise the motion of the system (3.112). At the same time, we have

$$\frac{dx}{dt} = (A(t) - BK_0C)x(t) - \sum_{i=1}^{m} G_i(t, x(t))K_iCx(t). \tag{3.115}$$

$\square$

It follows from the relation (3.115) that

$$x(t) = \Phi(t)\Phi^{-1}(t_0)x_0 - \int_{t_0}^{t} \Phi(t)\Phi^{-1}(s) \sum_{i=1}^{m} G_i(s, x(s))K_iCx(s)\, ds \tag{3.116}$$

where $x_0 \in [\underline{x}_0, \overline{x}_0]$. Taking into account the conditions of Theorem 3.15, from the relation (3.116), we obtain an estimate for the norm of the solution of the system (3.112) in the form

$$\|x(t)\| \leq \|x_0\|_2 Me^{-\alpha t} + \int_0^t \gamma Me^{-\alpha(t-s)} \sum_{i=1}^{m} \|K_iC\| \|x(s)\|^{q+1}\, ds. \tag{3.117}$$

Let us transform the inequality (3.117) to the following:

$$\|x(t)e^{\alpha t}\| \leq \|x_0\|_2 M + \int_0^t \gamma Me^{-\alpha qs} \sum_{i=1}^{m} \|K_iC\| \|x(s)e^{\alpha s}\|^{q+1}\, ds. \tag{3.118}$$

Applying Lemma 3.6 to the inequality (3.118), we obtain

$$\|x(t)e^{\alpha t}\| \leq \frac{M\|x_0\|_2}{\left(1 - \gamma q M^{q+1}\left(\sum_{i=1}^{m}\|K_iC\|\right)\|x_0\|_2^q \int_0^t e^{-\alpha q s}\,ds\right)^{\frac{1}{q}}}$$

$$= \frac{M\|x_0\|_2}{\left(1 + \frac{\gamma M^{q+1}\left(\sum_{i=1}^{m}\|K_iC\|\right)\|x_0\|_2^q}{\alpha}(e^{-\alpha q t} - 1)\right)^{\frac{1}{q}}}$$

for all $t \geq 0$.

Assume that the condition

$$\frac{\gamma M^{q+1}\left(\sum_{i=1}^{m}\|K_iC\|\right)\|x_0\|_2^q}{\alpha} < 1 \tag{3.119}$$

is satisfied. Then, for all $t \in [0, +\infty]$, the estimate

$$\|x(t)e^{\alpha t}\| \leq \frac{M\|x_0\|_2}{\left(1 - \frac{\gamma M^{q+1}\left(\sum_{i=1}^{m}\|K_iC\|\right)\|x_0\|_2^q}{\alpha}\right)^{\frac{1}{q}}}$$

holds, and for the norm of the solution $x(t)$, we obtain the estimate

$$\|x(t)\| \leq M_0\|x_0\|_2 e^{-\alpha t}$$

for all $t \geq 0$, where

$$M_0 = \frac{M}{\left(1 - \frac{\gamma M^{q+1}\left(\sum_{i=1}^{m}\|K_iC\|\right)\|x_0\|_2^q}{\alpha}\right)^{\frac{1}{q}}}.$$

Theorem 3.15 is proved.

3.5.4   Applications

As an illustration of the results obtained in the previous section, consider a mechanical system consisting of two *mathematical pendulums* of length l, each connected by a spring at a distance a from the suspension points.

It is assumed that their suspension points are at the same height. The pendulums are controlled by two oppositely directed forces v_1 and v_2, which are applied to the pendulum weights of mass m each. In addition, each of the pendulums is subject to the action of controlled generalised forces $Q_1 = -u_1\theta_1^2$ and $Q_2 = -u_2\theta_2^2$ (which can be interpreted as, for example, friction forces arising due to the oscillation of the pendulum in a liquid medium), where θ_1 and θ_2 are the angles that the pendulums form with the vertical.

The equations of motion of the proposed mechanical system can be written as

$$\begin{cases} ml^2\ddot{\theta}_1 + ka^2(\theta_1 - \theta_2) + mgl\theta_1 = -v_1 - u_1\theta_1^2, \\ ml^2\ddot{\theta}_2 + ka^2(\theta_2 - \theta_1) + mgl\theta_2 = v_2 - u_2\theta_2^2. \end{cases} \quad (3.120)$$

We also assume that the initial state of the system is determined by some interval vector

$$\Theta_0 = (\theta_{10}, \dot{\theta}_{10}, \theta_{20}, \dot{\theta}_{20})^T \in [\underline{\Theta}_0, \overline{\Theta}_0]. \quad (3.121)$$

Let us set the problem to find out what the controls $u = (u_1, u_2)^T$ and $v = (v_1, v_2)^T$ should be in order to stabilise the motion of the system (3.119) with the interval initial values (3.121) to an exponentially stable one.

Let us write the system (3.120) in normal form; for this, we introduce new variables x_1, x_2, x_3, and x_4 as follows:

$$x_1 = \theta_1,\ x_2 = \dot{\theta}_1,\ x_3 = \theta_2,\ x_4 = \dot{\theta}_2. \quad (3.122)$$

As a result, we obtain a system of equations,

$$\begin{cases} \dot{x}_1 = x_2, \\ \dot{x}_2 = \left(-\dfrac{ka^2}{ml^2} - \dfrac{g}{l}\right)x_1 + \left(-\dfrac{ka^2}{ml^2}\right)x_3 - \dfrac{u_1}{ml^2}x_1^2 - \dfrac{v_1}{ml^2}, \\ \dot{x}_3 = x_4, \\ \dot{x}_4 = \left(\dfrac{ka^2}{ml^2}\right)x_1 + \left(-\dfrac{ka^2}{ml^2} - \dfrac{g}{l}\right)x_3 - \dfrac{u_2}{ml^2}x_3^2 + \dfrac{v_2}{ml^2}, \end{cases} \quad (3.123)$$

with the interval initial conditions

$$x_i(t_0) = \Theta_0 \in [\underline{\Theta}_0, \overline{\Theta}_0],\ i = 1, 2, 3, 4. \quad (3.124)$$

The system (3.123) can also be represented as

$$\begin{cases} \dot{x}_1 = x_2, \\ \dot{x}_2 = px_1 - qx_3 - ru_1x_1^2 - rv_1, \\ \dot{x}_3 = x_4, \\ \dot{x}_4 = qx_1 + px_3 - en_2x_3^2 + rv_2, \end{cases} \tag{3.125}$$

where $p = -\frac{ka^2}{ml^2} - \frac{g}{l}$, $q = \frac{ka^2}{ml^2}$, and $r = \frac{1}{ml^2}$.

Using the notation of Section 3.4.3 of this chapter, we rewrite the system (3.125) as follows:

$$\frac{dx}{dt} = Ax + G(x)u + Bv, \tag{3.126}$$

where

$$A = \begin{bmatrix} 0 & 1 & 0 & 0 \\ p & 0 & -q & 0 \\ 0 & 0 & 0 & 1 \\ q & 0 & p & 0 \end{bmatrix}, \quad G(x) = \begin{bmatrix} 0 & 0 \\ -rx_1^2 & 0 \\ 0 & 0 \\ 0 & -rx_3^2 \end{bmatrix}, \quad B = \begin{bmatrix} 0 & 0 \\ -r & 0 \\ 0 & 0 \\ 0 & r \end{bmatrix},$$

$$u = (u_1, u_2)^T, \quad v = (v_1, v_2)^T.$$

The output vector y is assumed equal to the phase vector x, that is, the matrix C is equal to the identity matrix.

Let us check if the system (3.126) satisfies assumptions C_3–C_5. Since $G(0) = 0$, the assumption C_3 is satisfied. Let us check condition C_4. As K_0, consider the matrix

$$K_0 = \begin{bmatrix} \frac{\alpha - p}{r} & \frac{\beta}{r} & \frac{q}{r} & 0 \\ \frac{q}{r} & 0 & \frac{p - \alpha}{r} & -\frac{\beta}{r} \end{bmatrix},$$

where α, β are negative real numbers. Then, $A - BK_0C$ is equal to

$$A - BK_0C = \begin{bmatrix} 0 & 1 & 0 & 0 \\ \alpha & \beta & 0 & 0 \\ 0 & 0 & 0 & 1 \\ 0 & 0 & \alpha & \beta \end{bmatrix}.$$

We find the eigenvalues of the matrix $A - BK_0C$. The characteristic equation has the form

$$(\lambda^2 - \beta\lambda - \alpha)^2 = 0.$$

Since the parameters α and β are negative, all eigenvalues have negative real parts, that is, there exists $\mu > 0$ such that

$$\max_{i=\overline{1,4}} \operatorname{Re} \lambda_i(A - BK_0C) = -\mu < 0. \tag{3.127}$$

Let us estimate the norm of the matrix $\Phi(t)\Phi^{-1}(s)$, where $\Phi(t)$ is the fundamental matrix of the system of differential equations

$$\frac{dy}{dt} = (A - BK_0C)y.$$

Since the matrix $A - BK_0C$ is constant, $\Phi(t) = e^{(A-BK_0C)t}$ and $\Phi(t)\Phi^{-1}(s) = e^{(A-BK_0C)(t-s)}$ for all $t \geq s \geq t_0$. Using the well-known bound for the matrix exponent

$$\|e^{At}\| \leq Me^{\chi t},$$

where $\chi = \max_i \operatorname{Re} \lambda_i(A)$, $t \geq 0$, and the equality (3.127), we get the estimate

$$\|\Phi(t)\Phi^{-1}(s)\| \leq Me^{-\mu(t-s)},$$

that is, assumption C_4 is satisfied. We finally verify assumption C_5. Let us estimate the norm of the matrix $G(x)$:

$$\|G(x)\| = \sqrt{(-rx_1^2)^2 + (-rx_3^2)^2} = r\sqrt{x_1^4 + x_3^4} \leq r\|x\|^2,$$

that is, assumption H_5 is also satisfied.

Thus, according to the control Theorem 3.15, $u = -Ky$ and $v = -K_0y$, where the matrix $K_0 = \begin{bmatrix} \frac{\alpha-p}{r} & \frac{\beta}{r} & \frac{q}{r} & 0 \\ \frac{q}{r} & 0 & \frac{p-\alpha}{r} & -\frac{\beta}{r} \end{bmatrix}$, α and β are negative real numbers, stabilise the motion of the system (3.120) with the interval initial conditions (3.121) to exponentially stable. At the same time, the matrix K satisfies the inequality (3.119), which, when using the notation of this problem, has the form

$$\|K\| \cdot \|\Theta_0\|_I^2 < \frac{\mu}{rM^3}, \tag{3.128}$$

where $\|\Theta_0\|_I = \max\{\|\Theta_0\| : \Theta_0 \in [\underline{\Theta}_0, \overline{\Theta}_0]\}$.

3.6 Comments and Bibliography

It is known that for a number of mechanical systems (e.g., low-orbit artificial satellites) and in problems of celestial mechanics (e.g., the three-body problem), systems of ordinary differential equations with a polynomial right-hand side are satisfactory models (see Aminov [6], Babadzhanyants [9], and the bibliography therein). A. Poincaré (see Poincaré [138], Chapters XVI and XVII) showed that any differential equation under certain conditions can be reduced to a polynomial form by introducing additional variables.

When studying polynomial systems by the direct Lyapunov method, the central problem involves estimating the variation of the auxiliary function along the solutions of the equations under consideration (see Sirazetdinov and Aminov [155]).

In Chapter 3, based on the pseudo-linear representation of the nonlinear integral inequality, new boundaries for the variation of the auxiliary Lyapunov function are obtained, and applications are given in the study of various types of motion stability.

Section 3.2 is based on the results of the paper by Martynyuk and Chernienko [102].

Section 3.3 adapts the results of the paper by Martynyuk and Chernienko [105].

The results of the paper by Martynyuk and Chernienko [104] are the basis of Section 3.4.

The dynamics of mechanical and other systems under interval initial conditions is formalised using interval analysis (see Alefeld and Mayer [5] and Shary [156]). Moore's monograph [123] contains the first results on the interval estimation of solutions to the Cauchy problem for ordinary differential equations. Numerical analysis of solutions to systems of differential equations under interval initial conditions has been the subject of many works (see Rogalev [147] and the bibliography therein).

Section 3.5 is based on the results of the paper by Babenko and Martynyuk [10].

The relationship between optimal control and dissipativity is studied in detail in the monograph by Brogliato, Lozano, Maschke, and Egeland [25]. In this context, it is of interest to apply the results obtained in this chapter to dissipative systems.

Chapter 4

Stabilisation of Motions of Interval Bilinear and Affine Systems

4.1 Introduction

This chapter discusses the problems of stabilising the motion of interval bilinear and affine systems based on the integral method of qualitative analysis.

The chapter is organised according to the following plan.

In Section 4.2, the problem of robust stability of a bilinear system is formulated under interval initial conditions.

Section 4.3 establishes sufficient conditions for the robust stability of a bilinear system with uncertain parameter values.

In Section 4.4, conditions for robust stabilisation of the observable system are found.

In Section 4.5, the results obtained are applied to the analysis of vehicle wheel braking.

In Section 4.6, we present a statement of the problem of stabilising solutions of an affine system under interval initial conditions.

In Section 4.7, sufficient conditions for the stability of motion on a finite interval of an uncertain affine system are obtained under interval initial conditions.

The final section contains some discussions of the results obtained and a bibliography of papers used in writing this chapter.

4.2   Formulation of the Problem

Consider the system of equations of controlled motion

$$\frac{dx}{dt} = f(t, x, u),\tag{4.1}$$

$$x(t_0) = x_0 \in [\underline{x}_0, \overline{x}_0],\tag{4.2}$$

where $x \in \mathbb{R}^n$, $u \in \mathbb{R}^m$, $f : \mathbb{R}_+ \times \mathbb{R}^n \times \mathbb{R}^m \to \mathbb{R}^n$, $f(t, 0, u) = 0$, for all $t \in \mathbb{R}_+$ and $u \in U_0 \subset \mathbb{R}^m$, with the interval initial conditions (4.2). In addition, it is assumed that $f^T(t, x, u) f(t, x, u) \leq \eta^2 x^T H^T H x$, where $\eta > 0$ characterises the uncertainty boundary and H is some constant matrix.

Next, we formulate the definition of robust exponential stability under interval initial conditions.

Definition 4.1. The solution $x = 0$ of the system (4.1) under some control $u = u_0(t) \in U_0$ is called *exponentially stable* under the interval initial conditions (4.2) if there exist positive constants M and α such that for any $t_0 \in \mathbb{R}_+$ and $x_0 \in [\underline{x}_0, \overline{x}_0]$, the estimate

$$\|x(t; t_0, x_0)\| \leq M \|x_0\|_I e^{-\alpha(t - t_0)}$$

is valid for all $t \geq t_0$.

Definition 4.2. The system (4.1) is *robustly exponentially stabilisable* under the interval initial conditions (4.2) if there exists a control $u = u_0(t) \in U_0$ for which the solution $x = 0$ of the system (4.1) is exponentially stable under the interval initial conditions (4.2).

4.3   Robust Stabilisation of a Bilinear System

We consider the equations of *controlled motion* of a *bilinear system* with uncertain values of the parameters

$$\frac{dx}{dt} = Ax(t) + \sum_{i=1}^{p} u_i(t) B_i x(t) + D u_0(t)\tag{4.3}$$

under the interval initial conditions (4.2).

Here, $x \in \mathbb{R}^n$, $A = A_0 + \Delta A$, $B_i = B_{i0} + \Delta B_i$, $D = D_0 + \Delta D$ are matrices of corresponding dimensions, the matrices A_0, B_{i0}, and D_0 correspond to the nominal system, and ΔA, ΔB_i, and ΔD are the uncertainties of the system parameters and control.

Note that for the fixed dimensions of the matrices $A = A_0 + \Delta A$, $B_i = B_{i0} + \Delta B_i$, and $D = D_0 + \Delta D$ in the system (4.3) in the control expressions $u_i(t)$ and $u_0(t)$, the matrices K_i and K_0 have the dimensions at which the right-hand side of the equations (4.3) has the dimension of the state vector systems (4.3).

We make the following assumptions about the system (4.3):

D_1. There are positive constants $a, b_i, c, i = 1, 2, \ldots, p$ such that $\|\Delta A\| \le a$, $\|\Delta B_i\| \le b_i$, and $\|\Delta D\| \le c$.

D_2. The *pair* (A_0, D_0) is *stabilisable*, that is, there exists a matrix K such that under the control $u_0(t) = -Kx(t)$, the matrix $\widetilde{A} = A_0 - D_0 K$ is such that

$$\|e^{\widetilde{A}t}\| \le M_0 e^{\omega t},$$

where $M_0 > 0$ and $\omega < 0$.

Let us rewrite the system (4.3) as

$$\frac{dx}{dt} = (A_0 - D_0 K)x(t) + \sum_{i=1}^{p} u_i(t)(B_{i0} + \Delta B_i)x(t) + (\Delta A - \Delta D K)x(t),$$

$$(4.4)$$

and suppose that the local controls

$$u_i(t) = -K_i x(t), \quad i = 1, 2, \ldots, p, \tag{4.5}$$

together with the global control

$$u_0(t) = -Kx(t), \tag{4.6}$$

implement the specified properties of the motion of the object, which is described by the system (4.4).

Let us indicate the conditions under which the controls (4.5) and (4.6) stabilise the motion of the system (4.4) to a robustly exponentially stable one.

The following assertion holds.

Theorem 4.1. *Assume that the system (4.4) satisfies the conditions of assumptions D_1 and D_2 and, in addition,*

$$\omega + M_0(a + c\|K\|) + \|x_0\|_I M_0^2 \sum_{i=1}^{p} \|K_i\|(\|B_{i0}\| + b_i) < 0. \tag{4.7}$$

Then, the controls $u_0(t)$ and $u_i(t)$ stabilise the motion of the system (4.4) to a robustly exponentially stable one.

Proof. Considering (4.7), from the equation (4.4), we find that

$$x(t) = e^{\widetilde{A}t}x_0 - \int_0^t e^{\widetilde{A}(t-s)} \sum_{i=1}^{p} K_i x(s)(B_{i0}$$

$$+ \Delta B_i)x(s)\, ds + \int_0^t e^{\widetilde{A}(t-s)}(\Delta A - \Delta DK)x(s)\, ds, \qquad (4.8)$$

where $x_0 \in [\underline{x}_0, \overline{x}_0]$. If condition D_2 is fulfilled for the system

$$\frac{d\widetilde{x}}{dt} = (A_0 - D_0 K)\widetilde{x}, \qquad (4.9)$$

there are constants $M_0 > 0$ and $\omega < 0$ such that

$$\|e^{\widetilde{A}t}\| \le M_0 e^{\omega t} \qquad (4.10)$$

for all $t \ge 0$. Taking into account the condition D_1 and the estimate (4.10), from the relation (4.8), we obtain the inequality

$$\psi(t) \le m_I + \int_0^t [f_1\psi(s) + f_2(s)\psi^2(s)]\, ds, \qquad (4.11)$$

where $\psi(t) = \|x(t)\|e^{-\omega t}$, $m_I = M_0\|x_0\|_I$, $\|x_0\|_I = \max\{\|x(t_0)\| : x(t_0) = x_0 \in [\underline{x}_0, \overline{x}_0]\}$, $f_1 = M_0(a + c\|K\|)$, and $f_2(s) = M_0 \sum_{i=1}^{p} \|K_i\| \times (\|B_{i0}\| + b_i)e^{\omega s}$. $\qquad \square$

Applying Lemma 1.2 to the inequality (4.11), we obtain the estimate

$$\psi(t) \le m_I \exp\left\{\int_0^t [f_1 + f_2(s)\psi(s)]\, ds\right\}. \qquad (4.12)$$

From (4.12), it follows that

$$-\psi(t)\exp\left\{\int_0^t (-f_2(s)\psi(s))\, ds\right\} \ge -m_I \exp\int_0^t f_1\, ds. \qquad (4.13)$$

Multiplying both sides of the inequality (4.13) by $f_2(t) > 0$, for all $t \ge 0$, we obtain

$$\frac{d}{dt}\left\{\exp\left[-\int_0^t f_2(s)\psi(s)\, ds\right]\right\} \ge -m_I f_2(t)\exp\int_0^t f_1\, ds.$$

Integrating this inequality between 0 and t, we find that

$$(\psi(t))^{-1}m_I \exp\int_0^t f_1\, ds \ge 1 - m_I \int_0^t f_2(s)\exp\left\{\int_0^s f_1\, d\eta\right\}ds. \qquad (4.14)$$

In order to resolve this inequality with respect to $\psi(t)$, we find the sign of the expression on the right-hand side of the inequality. To this end, we simplify it by calculating the integrals included in it. As a result of simple transformations, we get

$$1 - m_I \int_0^t f_2(s) \exp\left\{\int_0^s f_1 \, d\eta\right\} ds$$

$$= 1 - \frac{m_I M_0 \sum\limits_{i=1}^p \|K_i\|(\|B_{i0}\| + b_i)}{\omega + M_0(a + c\|K\|)} \left[e^{(\omega + M_0(a+c\|K\|))t} - 1\right]. \quad (4.15)$$

Let us introduce the notation $\omega_1 = \omega + M_0(a + c\|K\|)$. Since, following from the inequality (4.7), $\omega_1 < 0$ and $e^{\omega_1 t} > 0$ for any $t \geq 0$ (and, moreover, the greater the accuracy of the last estimate, the larger the value of t), we get the following estimate:

$$1 - m_I \int_0^t f_2(s) \exp\left\{\int_0^s f_1 \, d\eta\right\} ds > 1 + \frac{m_I M_0 \sum\limits_{i=1}^p \|K_i\|(\|B_{i0}\| + b_i)}{\omega_1},$$

which is valid for any $t \geq 0$. The last expression is positive if we use the condition (4.7) of Theorem 4.1 and take into account the notation for the constant m_I. Thus, the right-hand side of the inequality (4.14), by virtue of the condition of Theorem 4.1, is positive for any $t \geq 0$; therefore, (4.14) can be solved for $\psi(t)$ as follows:

$$\psi(t) \leq \frac{m_I \exp \int_0^t f_1 \, ds}{1 - m_I \int_0^t f_2(s) \exp \int_0^s f_1 \, d\eta \, ds}$$

$$= \frac{m_I e^{M_0(a+c\|K\|)t}}{1 - \frac{m_I M_0 \sum\limits_{i=1}^p \|K_i\|(\|B_{i0}\| + b_i)}{\omega_1} \left(e^{\omega_1 t} - 1\right)}. \quad (4.16)$$

Using again the estimate $e^{\omega_1 t} > 0$, which is valid for all $t \geq 0$, we obtain the following estimate for $\psi(t)$:

$$\psi(t) \leq M(m_I) e^{M_0(a+c\|K\|)t} \quad (4.17)$$

for all $t \geq 0$, where

$$M(u) = \frac{u\omega_1}{\omega_1 + u M_0 \sum\limits_{i=1}^p \|K_i\|(\|B_{i0}\| + b_i)}.$$

Considering that $\psi(t) = \|x(t)\|e^{-\omega t}$, from (4.17), we obtain an estimate for $\|x(t)\|$:

$$\|x(t)\| \le M(M_0\|x_0\|_I)e^{\omega_1 t} \tag{4.18}$$

for all $t \ge 0$.

This proves the assertion of Theorem 4.1.

4.4 Robust Stabilisation of the Observable System

Further, we assume that in the system (4.3), $p = 1$ and consider the system of equations

$$\frac{dx}{dt} = Ax(t) + u(t)Bx(t) + Du_0(t),$$
$$y(t) = Cx(t), \tag{4.19}$$

$$x(t_0) = x_0 \in [\underline{x}_0, \overline{x}_0], \tag{4.20}$$

where $A = A_0 + \Delta A$, $B = B_0 + \Delta B$, and $D = D_0 + \Delta D$ are matrices of corresponding dimensions. Regarding the system (4.19), suppose the following:

D_3. The *pair* (A_0, C) is *observable*, i.e., there exists a matrix G such that the zero solution of the system

$$\frac{d\overline{x}}{dt} = (A_0 - GC)\overline{x} \tag{4.21}$$

is asymptotically stable and, therefore, there are a matrix K and constants $N_0 > 0$ and $\lambda < 0$ for which the estimate

$$\|e^{(A_0 - GC)t}\| \le N_0 e^{\lambda t} \tag{4.22}$$

holds for all $t \ge 0$.

D_4. The *pair* (A_0, D_0) *is stabilisable* if there exist constants $N_1 > 0$ and $\beta < 0$ such that the inequality

$$\|e^{(A_0 - D_0 K)t}\| \le N_1 e^{\beta t}$$

holds for all $t \ge 0$.

Following the general theory of observation (see [122]), suppose that the "observer" in the system (4.19) is described by the system of equations

$$\frac{dz}{dt} = \Phi z(t) + u(t)B_0 z(t) + D_0 u(t) + Gy(t), \tag{4.23}$$

$$z(t_0) = z_0 \in [\underline{z}_0, \overline{z}_0], \tag{4.24}$$

where $\Phi = A_0 - GC$ is an $n \times n$ constant matrix.

Let us introduce the extended vector of the system (4.19) and the *observation errors* $\varepsilon(t) = z(t) - x(t)$ in the form

$$w(t) = (x(t), \varepsilon(t))^T, \quad w(0) = w_0 \in [\underline{w}_0, \overline{w}_0],$$

and we rewrite the equation (4.19) together with the error equation as

$$\frac{dx}{dt} = \widetilde{A}x(t) + (\Delta A - \Delta DK)x(t) - Kz(t)Bx(t) - DK\varepsilon(t),$$

$$\frac{d\varepsilon}{dt} = \Phi\varepsilon(t) - (\Delta A + \Delta DK)x(t) - Kz(t)(B_0\varepsilon(t) - \Delta Bx(t)),$$

(4.25)

where $\widetilde{A} = A_0 - D_0K$ and $\Phi = A_0 - GC$. This takes into account the fact that the control $u(t) = -Kz(t)$.

The system (4.25), taking into account the content of the vector $w(t)$ and the right-hand side, is rewritten as

$$\frac{dw}{dt} = Hw(t) + Y(t) + Z(x(t)),$$

$$w(0) = w_0 \in [\underline{w}_0, \overline{w}_0],$$

(4.26)

where $H = \begin{pmatrix} \widetilde{A} & -D_0K \\ 0 & \Phi \end{pmatrix}$, $Y(t) = \begin{pmatrix} -Kz(t)(B_0 + \Delta B)x(t) \\ -Kz(t)(B_0\varepsilon - \Delta Bx(t)) \end{pmatrix}$, and

$Z(x(t)) = \begin{pmatrix} (\Delta A - \Delta DK)x(t) \\ -(\Delta A + \Delta DK)x(t) \end{pmatrix}$.

From the equation (4.26), we find that

$$w(t) = e^{Ht}w_0 + \int_0^t e^{H(t-s)}(Z(x(s)) + Y(s))\, ds. \qquad (4.27)$$

According to Appendix B, we have the estimate

$$\|e^{Ht}\| \le \widetilde{N}e^{\sigma t}, \quad \sigma = \max(\lambda, \beta), \qquad (4.28)$$

where $\widetilde{N} = const > 0$ and, moreover,

$$\|Z(x(t))\| \le 2(a + c\|K\|)\|w(t)\|, \quad \|Y(t)\| \le 4\|K\,|(\|B_0\| + b)\|w(t)\|^2 \qquad (4.29)$$

for all $t \ge 0$.

Next, denote $\widetilde{f}_1 = 2\widetilde{N}(a + c\|K\|)$ and $\widetilde{f}_2(s) = 4\widetilde{N}\|K\|(\|B_0\| + b)e^{\sigma s}$, and obtain from the relation (4.27) the estimate

$$\widetilde{\psi}(t) \le \widetilde{m}_I + \int_0^t \left(\widetilde{f}_1\widetilde{\psi}(s) + \widetilde{f}_2(s)\widetilde{\psi}^2(s) \right) ds, \qquad (4.30)$$

where $\widetilde{\psi}(t) = \|w(t)\|e^{-\sigma t}$, $\widetilde{m}_I = \widetilde{N}\|w_0\|_I$, $\|w_0\|_I = \max\{\|w(t_0)\| : w(t_0) = w_0 \in [\underline{w}_0, \overline{w}_0]\}$.

Let us apply the estimation technique from Lemma 1.4 to the inequality (4.30) and obtain the inequality

$$\widetilde{\psi}(t) \leq \frac{\widetilde{m}_I \exp \int_0^t \widetilde{f}_1 \, ds}{1 - \widetilde{m}_I \int_0^t \widetilde{f}_2(s) \exp \int_0^s \widetilde{f}_1(\eta) \, d\eta \, ds}, \tag{4.31}$$

which is true for all $t \geq 0$ unless

$$\sigma + 2\widetilde{N}(a + c\|K\|) + 4\|w_0\|_I \widetilde{N}^2 \|K\|(\|B_0\| + b) < 0. \tag{4.32}$$

The following assertion holds.

Theorem 4.2. *Let the system* (4.19) *satisfy the assumptions* D_1 *and* D_2, *and in addition, let the condition* (4.32) *be satisfied. Then, the system* (4.19) *is robust exponentially stabilisable with the estimate*

$$\|w(t)\| \leq \widetilde{M}(\widetilde{N}\|w_0\|_I) e^{\widetilde{\omega}_1 t}, \tag{4.33}$$

where $\widetilde{\omega}_1 = \sigma + 2\widetilde{N}(a + c\|K\|) < 0$, $\sigma = \max(\alpha, \beta)$ *and*

$$\widetilde{M}(u) = \frac{u\widetilde{\omega}_1}{\widetilde{\omega}_1 + 4u\widetilde{N}\|K\|(\|B_0\| + b)}.$$

Proof. Under the conditions D_1 and D_2, we obtain the estimate (4.31), which holds for all $t \in [0, \infty)$ if the condition (4.32) holds. From (4.31), taking into account the notation $\widetilde{\psi}$, $\widetilde{f}_1$, and $\widetilde{f}_2(s)$, we find the expression for the constant $\widetilde{M}$ in the inequality (4.33), from which, under the condition (4.32), the assertion of Theorem 4.2) follows.   □

4.5   Applications

We study the dynamics of car braking under the following assumptions (see [122]). Consider a car wheel that rotates on an axle with a certain angular velocity ω. The braking system consists of a drum that is stationary relative to the wheel and a brake shoe that leans against the drum with different densities, thereby creating a friction force f_1 of different intensities.

Suppose that:

(1) $F_1 = c_1 u_1 \dot{x}$, where the control u_1 is a normal force applied to block, $\dot{x}$ is a translational speed, and c_1 is a constant;
(2) *Coulomb friction* is absent;
(3) other *friction forces* F_2 are expressed as $F_2 = c_2 \dot{x}$, where c_2 is a constant.

Taking into account the forces of inertia, friction, and thrust of the engine, according to Newton's second law, we obtain the equation of motion in the form

$$m\frac{d}{dt}(\dot{x}) = -kc_1u_1\dot{x} - kc_2\dot{x} + u, \qquad (4.34)$$

where k is the proportionality factor, m is the mass of the car, and u is the force due to the engine. Denoting $y_1 = x$ and $y_2 = \dot{x}$, the equation (4.34) can be rewritten as a system,

$$\begin{cases} \dfrac{dy_1}{dt} = y_2, \\[2mm] \dfrac{dy_2}{dt} = \left(-\dfrac{c_1u_1}{m} - \dfrac{c_2}{m}\right)y_2 + \dfrac{u}{m}, \end{cases}$$

or in the vector form

$$\frac{dy}{dt} = Ay + u_1By + Du, \qquad (4.35)$$

where

$$y = \begin{pmatrix} y_1 \\ y_2 \end{pmatrix}, \quad A = \begin{pmatrix} 0 & 1 \\ 0 & -\frac{kc_2}{m} \end{pmatrix}, \quad B = \begin{pmatrix} 0 & 0 \\ 0 & -\frac{kc_1}{m} \end{pmatrix}, \quad D = \begin{pmatrix} 0 \\ \frac{1}{m} \end{pmatrix}.$$

Assume that the parameters m, c_2, and c_1 are given imprecisely, and take values from the bounded intervals $[m_0 - \Delta m, m_0 + \Delta m]$, $[c_{20} - \Delta c_2, c_{20} + \Delta c_2]$, and $[c_{10} - \Delta c_1, c_{10} + \Delta c_1]$, respectively. Then, A, B, and C can be represented as $A = A_0 + \Delta A$, $B = B_0 + \Delta B$, and $D = D_0 + \Delta D$, where

$$A_0 = \begin{pmatrix} 0 & 1 \\ 0 & -\frac{kc_{20}}{m_0} \end{pmatrix}, \quad B_0 = \begin{pmatrix} 0 & 0 \\ 0 & -\frac{kc_{10}}{m_0} \end{pmatrix}, \quad D_0 = \begin{pmatrix} 0 \\ \frac{1}{m_0} \end{pmatrix},$$

and ΔA, ΔB, and ΔC are gains due to parameter uncertainty.

Let us find the conditions that the matrices $K = [k_1, k_2]$ and $K_1 = [k_{11}, k_{12}]$ must satisfy so that the controls $u_1 = -K_1y$ and $u = -Ky$ robustly exponentially stabilise the car wheel. To do this, we first find out what the matrix K should be for the pair (A_0, D_0) to be stabilisable, that is, there are numbers $M_0 > 0$ and $\omega < 0$ such that

$$\|e^{(A_0-D_0K)t}\| \le M_0e^{\omega t} \qquad (4.36)$$

for all $t \ge 0$, that is, for hypothesis H_2 to hold.

We choose the matrix K by requiring the fulfilment of the condition

$$\max_{i=1,2} \operatorname{Re} \lambda_i(A_0 - D_0K) < 0.$$

This happens when

$$(A_0 - D_0K) < 0 \quad \text{and} \quad \det(A_0 - D_0K) > 0. \tag{4.37}$$

Because $A_0 - D_0K = \begin{pmatrix} 0 & 1 \\ -\frac{k_1}{m_0} & \frac{-kc_{20}-k_2}{m_0} \end{pmatrix}$, the conditions (4.37) become

$$\begin{cases} k_2 > -kc_{20}, \\ k_1 > 0. \end{cases} \tag{4.38}$$

Denoting $\alpha_0 = \max_{i=1,2} \operatorname{Re} \lambda_i(A_0 - D_0K)$ and using Lemma B.1 (see Appendix B), we obtain the estimate (4.36), where M_0 and ω are found using the formulas

$$M_0 = \frac{2\|A_0 - D_0K\|}{\alpha} e^{-1 + \frac{\alpha}{2\|A_0 - D_0K\|}}, \quad \omega = \alpha_0 + \alpha, \tag{4.39}$$

where $\alpha > 0$ is chosen so that $\alpha_0 + \alpha < 0$.

In order to obtain conditions for the robust exponential stability of the solution $y = 0$ of the system (4.35), we additionally require the fulfilment of the inequality (4.7) from the conditions of Theorem 4.1:

$$\omega + M_0(a + c\|K\|) + \|y_0\|_I M_0^2 \sum_{i=1}^{p} \|K_i\|(\|B_{i0}\| + b_i) < 0.$$

In this case, the sum on the left-hand side of the inequality contains only one term, so the inequality becomes

$$\omega + M_0(a + c\|K\|) + \|y_0\|_I M_0^2 \|K_1\|(\|B_0\| + b) < 0, \tag{4.40}$$

where a, b, and c are positive numbers such that $\|\Delta A\| \le a$, $\|\Delta B\| \le b$, and $\|\Delta D\| \le c$. Since $\|B_0\| = \frac{kc_{10}}{m_0}$, $\|K\| = \sqrt{k_1^2 + k_2^2}$, and $\|K_1\| = \sqrt{k_{11}^2 + k_{12}^2}$, the inequality (4.40) can be rewritten as follows:

$$\omega + M_0\left(a + c\sqrt{k_1^2 + k_2^2}\right) + \|y_0\|_I M_0^2 \sqrt{k_{11}^2 + k_{12}^2}\left(\frac{kc_{10}}{m_0} + b\right) < 0. \tag{4.41}$$

Thus, if in the system (4.35), the controls u_1 and u have the form

$$u_1 = -K_1 y, \quad u = -Ky,$$

and the matrices $K_1 = [k_{11}, k_{12}]$, and $K = [k_1, k_2]$ satisfy the conditions (4.38) and (4.41), in which M_0 and ω are found using the formulas (4.39), where $\alpha > 0$ is chosen so that $\alpha_0 + \alpha < 0$ and a, b, and c are positive numbers such that $\|\Delta A\| \le a$, $\|\Delta B\| \le b$, and $\|\Delta D\| \le c$, then the solution $y = 0$ of the system (4.35) is robustly exponentially stable.

These are the sufficient conditions for robust exponential stabilisation of the car motion.

4.6   Stability and Stabilisation of Nonlinear Systems

In this section, we obtain new estimates for the norms of solutions of nonlinear systems and sufficient conditions for stabilising the motion of an affine system under interval initial conditions. These conditions are obtained on the basis of estimates of the norms of solutions of the corresponding systems of equations of perturbed motion. The proposed method of stabilisation is used in the study of an electromechanical system with a permanent magnet.

4.6.1   Estimates of the norms of solutions

Let the system of differential equations of perturbed motion be given as

$$\frac{dx}{dt} = A(t)x + g_1(t, x) + g_2(t, x), \tag{4.42}$$

$$x(t_0) = x_0 \in [\underline{x}_0, \overline{x}_0], \tag{4.43}$$

under the interval initial conditions $x(t_0) = x_0$ for the solution $x(t, t_0, x_0)$. Here, $x \in \mathbb{R}^n$ and $t \in \mathbb{R}_+$, and the vector functions g_1 and g_2 are defined and continuous on the product $\mathbb{R}_+ \times \mathbb{R}^n$. It is assumed that the vector functions g_1 and g_2 are sufficiently smooth and the solutions of the initial problem of equations (4.42) exist for all $t \geq t_0$.

Of interest is the answer to the following question: under what restrictions on the functions g_1 and g_2 is it possible to obtain a bound on the variation of the norm of the solution of the system (4.42), suitable for solving the problem of stabilising the motion of an affine system? To solve the problem, the method of nonlinear integral inequalities is used.

Regarding the system (4.42), suppose the following:

D_5. For all $t \geq t_0$, $A(t) = 0$.

D_6. There are integrable non-negative functions $h_1(t)$ and $h_2(t)$ and constants $p > 1$ and $q \geq 1$ such that for all $(t, x) \in \mathbb{R}_+ \times \mathbb{R}^n$,

$$\|g_1(t, x)\| \leq h_1(t)\|x\|^p,$$

$$\|g_2(t, x)\| \leq h_2(t)\|x\|^q.$$

Under conditions D_5 and D_6, from the relation

$$x(t) = x_0 + \int_{t_0}^{t} (g_1(s, x(s)) + g_2(s, x(s)))\, ds, \tag{4.44}$$

we get the inequality

$$\|x(t)\| \leq \|x_0\|_I + \int_{t_0}^{t} \left(h_1(s)\|x(s)\|^p + h_2(s)\|x(s)\|^q \right) ds$$

for all $t \geq t_0$, from which, taking into account the relationship between the interval and Euclidean norms (see Appendix A), we obtain the inequality

$$\|x(t)\| \leq \|x_0\|_I + \int_{t_0}^{t} \left(h_1(s)\|x(s)\|^p + h_2(s)\|x(s)\|^q \right) ds \qquad (4.45)$$

for all $t \geq t_0$.

Let us show that the following assertion holds.

Lemma 4.1. *Assume that the system* (4.42) *satisfies the conditions of assumptions D_5 and D_6 and*

$$1 - (p + q - 2) \left(\|x_0\|_I^{p-1} \int_{t_0}^{t} h_1(s)ds + \|x_0\|_I^{q-1} \int_{t_0}^{t} h_2(s)ds \right) > 0 \quad (4.46)$$

for all $t \in J$.

Then, the norm of the solution $x(t)$ of the system (4.42) *satisfies the estimate*

$$\|x(t)\|$$

$$\leq \frac{\|x_0\|_I}{\left[1 - (p + q - 2) \left(\|x_0\|_I^{p-1} \int_{t_0}^{t} h_1(s)ds + \|x_0\|_I^{q-1} \int_{t_0}^{t} h_2(s)ds \right) \right]^{\frac{1}{p+q-2}}}$$

$$(4.47)$$

for all $t \in J$.

Proof. Under conditions D_5 and D_6, it is easy to obtain the inequality (4.45). Let us rewrite this inequality in a pseudo-linear form:

$$\|x(t)\| \leq \|x_0\|_I + \int_{t_0}^{t} \left(h_1(s)\|x(s)\|^{p-1} + h_2(s)\|x(s)\|^{q-1} \right) \|x(s)\| ds,$$

and applying Lemma 1.2, we obtain the estimate

$$\|x(t)\| \leq \|x_0\|_I \exp \left[\int_{t_0}^{t} \left(h_1(s)\|x(s)\|^{p-1} + h_2(s)\|x(s)\|^{q-1} \right) ds \right]$$

for all $t \in J$. □

Further, the proof of Lemma 4.1 is similar to that of Lemma 1.7, and for this reason, we omit it here.

4.6.2   Stabilisation of the motion of an affine system

Consider a nonlinear *non-autonomous affine system* with many controls of the form (cf. [71, 124, 149]):

$$dx/dt = A(t)x(t) + Y(t,x) + \sum_{i=1}^{m} G_i(t, x(t))u_i(t) + Bu_0(t), \quad (4.48)$$

$$y(t) = Cx(t), \quad (4.49)$$

$$x(t_0) = x_0 \in [\underline{x}_0, \overline{x}_0]. \quad (4.50)$$

Here, the vector function $Y(t,x)$ is such that the set of solutions of the system (4.48) under the interval initial conditions (4.50) exists for all $t \in [t_0, a)$, $a = \text{const} > 0$; $x \in \mathbb{R}^n$, $A(t)$ is an $n \times n$ matrix with continuous entries on any finite interval of motion, $G_i(t,x)$ is an $n \times m$ matrix; the control vectors $u_i(t) \in \mathbb{R}^m$ for all $i = 1, 2, \ldots, m$; B is an $n \times m$ matrix; the control $u_0(t) \in \mathbb{R}^m$; C is a constant $n \times n$ matrix; and x_0 is an interval vector of the initial states of the system (4.50). We make the following assumptions about the system (4.48):

D_7. Functions $G_i(t, 0) = 0$, $i = 1, 2, \ldots, m$, for all $t \geq 0$.

D_8. There exists a constant $n \times m$ matrix K_0 such that for the system

$$\frac{dy}{dt} = (A(t) - BK_0C)y,$$

the fundamental matrix $\Phi(t)$ satisfies the estimate

$$\|\Phi(t)\Phi^{-1}(s)\| \leq Me^{-\alpha(t-s)} \quad (4.51)$$

for $t \geq s \geq t_0$, where M and α are some positive constants.

D_9. There are constants $\gamma_i > 0$ and $q \geq 1$ such that

$$\|G_i(t,x)\| \leq \gamma_i \|x\|^q \quad (4.52)$$

for all $i = 1, 2, \ldots, m$.

D_{10}. There are constants $\varkappa > 0$ and $p > 0$ such that

$$\|Y(t,x)\| \leq \varkappa \|x\|^{p+1}$$

in the range $(t,x) \in \mathbb{R}_+ \times D$.

Let us show that the following assertion is true.

Theorem 4.3. *Let the conditions of assumptions D_7–D_{10} be satisfied. Then, the controls*

$$u_i(t) = -K_i y(t), \quad i = 1, 2, \ldots, m, \quad u_0(t) = -K_0 y(t) \quad (4.53)$$

stabilise the motion $x = 0$ of the system (4.48) *with an estimate for the norm of solutions in the form*

$$\|x(t)\| \le M\|x_0\|_I e^{-\alpha t}$$

$$\times \left\{ 1 - (p+q) \left[\frac{\gamma \sum_{i=1}^{m} \|K_i C\| M^{q+1} \|x_0\|_I^q}{q} + \frac{\varkappa M^{p+1}\|x_0\|_I^p}{p} \right] \right\}^{-\frac{1}{p+q}}$$

for all $t \in [0, \sigma)$, where

$$\sigma = \sup \left\{ t \in \mathbb{R}_+ : (p+q) \int_0^t \left(\varkappa \|x_0\|_I^p M^{p+1} e^{p\alpha s} \right. \right.$$

$$\left. \left. + \gamma \sum_{i=1}^{m} \|K_i C\| M^{q+1} \|x_0\|_I^q e^{q\alpha s} \right) ds < 1 \right\}.$$

Proof. Let $m < n$. To stabilise the set of trajectories of the system (4.48) under the initial conditions (4.50), the controls are

$$u_i(t) = -K_i y(t), \quad u_0(t) = -K_0 y(t), \quad i = 1, 2, \ldots, m. \tag{4.54}$$

$\square$

From (4.48), under the controls (4.54), it follows that

$$dx/dt = (A(t) - BK_0 C)x(t) + Y(t, x(t)) + \sum_{i=1}^{m} G_i(t, x(t))(K_i C x(t)). \tag{4.55}$$

Under the condition D_{10}, from the relation (4.55), we obtain

$$x(t) = \Phi(t)\Phi^{-1}(t_0)x_0 + \int_{t_0}^t \Phi(t)\Phi^{-1}(t_0)Y(s, x(s))ds$$

$$- \int_{t_0}^t \Phi(t)\Phi^{-1}(t_0) \sum_{i=1}^{m} G_i(s, x(s))(K_i C x(s))ds, \tag{4.56}$$

where $x_0 \in [\underline{x}_0, \overline{x}_0]$. In view of the estimate (4.51), it follows from the relation (4.56) that

$$\|x(t)\| \le M e^{-\alpha t}\|x_0\|_I$$

$$+ M e^{-\alpha t} \int_0^t \left(\varkappa e^{\alpha s}\|x(s)\|^{p+1} + \gamma \sum_{i=1}^{m} e^{\alpha s}\|K_i C\|\|x(s)\|^{q+1} ds \right), \tag{4.57}$$

where $\gamma = \sum_{i=1}^{m} \gamma_i$. Further, the inequality (4.57) is transformed into the following:

$$\|x(t)\|e^{\alpha t} \leq M\|x_0\|_I + M \int_0^t \left(\varkappa e^{-p\alpha s}(e^{\alpha s}\|x(s)\|)^{p+1} \right.$$

$$\left. +\gamma \sum_{i=1}^{m} \|K_i C\|e^{-q\alpha s}(e^{\alpha s}\|x(s)\|)^{q+1} \right) ds. \qquad (4.58)$$

Applying Lemma 4.1 to the inequality (4.58), we obtain the estimate

$$\|x(t)\|$$

$$\leq M\|x_0\|_I e^{-\alpha t} \times \left[1 - (p+q) \left(M^{p+1}\varkappa\|x_0\|_I^p \int_0^t e^{-p\alpha s}ds \right.\right.$$

$$\left.\left. + M^{q+1}\gamma\|x_0\|_I^q \sum_{i=1}^{m} \|K_i C\| \int_0^t e^{-q\alpha s}ds \right) \right]^{-\frac{1}{p+q}},$$

which implies the assertion of Theorem 4.3.

4.6.3   Applications

As an example, consider the mathematical model of a permanent *magnet synchronous motor* (see [163]):

$$\begin{cases} L_d \dfrac{di_d}{dt} = -R_s i_d + n_p\omega L_q i_q + u_d, \\[2mm] L_q \dfrac{di_q}{dt} = -R_s i_q - n_p\omega L_d i_d - n_p\omega\Phi + u_q, \\[2mm] J \dfrac{d\omega}{dt} = \frac{3}{2}n_p\left[(L_d - L_q)i_d i_q + \Phi i_q\right] - \tau_L, \end{cases} \qquad (4.59)$$

where d and q denote the axes of the coordinate system rigidly connected to the rotor, i_d and i_q are the currents in the rotor windings, u_d and u_q are the voltages in the rotor windings, ω is the angular velocity of the rotor, J is the moment of inertia, L_d, L_q, R_s, n_p, Φ, and τ_l are positive constants. The parameters u_d, u_q, and J are the controls in the system (4.59).

Let us assume that for some values of the parameters of the controls $u_d = u_{d0}$, $u_q = u_{q0}$, and $J = J_0$, the system (4.59) has the equilibrium state

$$i_d = i_{d0},\ i_q = i_{q0},\ \omega = \omega_0. \qquad (4.60)$$

Let us set the problem of determining such a control law for an object, in which the equilibrium state (4.60) is stable. We solve the problem of stabilisation using linear feedback.

In the system (4.59), we perform a change of variables according to the formulas

$$i_d = i_{d0} + x_1,\ i_q = i_{q0} + x_2,\ \omega = \omega_0 + x_3,$$
$$u_d = u_{d0} + v_{01},\ u_q = u_{q0} + v_{02},\ J = J_0 + u. \tag{4.61}$$

Then, the system (4.59) will take the form

$$\frac{dx}{dt} = Ax(t) + Y(x(t)) + G(x(t))u(t) + Bv_0, \tag{4.62}$$

where

$$A = \begin{pmatrix} -\frac{R_s}{L_d} & \frac{n_p L_q \omega_0}{L_d} & \frac{n_p L_q i_{q0}}{L_d} \\ -\frac{n_p L_d \omega_0}{L_q} & -\frac{R_s}{L_q} & \left(-\frac{n_p L_d i_{d0}}{L_q} - n_p \Phi\right) \\ \frac{3n_p i_{d0}}{2J_0}(L_d - L_q) & \frac{3n_p i_{d0}}{2J_0}(L_d - L_q) + \frac{3n_p \Phi}{2} & 0 \end{pmatrix},$$

$$Y(x(t)) = \begin{pmatrix} \frac{n_p L_q}{L_d} x_2 x_3 \\ -\frac{n_p L_d}{L_q} x_1 x_3 \\ \frac{3n_p}{2J_0}(L_d - L_q)x_1 x_2 \end{pmatrix},$$

$$G(x(t)) = \left(0,\quad 0,\quad \frac{3n_p i_{q0}}{2J_0}(L_d - L_q)x_1 \right.$$
$$+ \left(\frac{3n_p i_{d0}}{2J_0}(L_d - L_q) + \frac{3n_p \Phi}{2}\right)x_2$$
$$\left. + \frac{3n_p}{2J_0}(L_d - L_q)x_1 x_2 \right)^T,$$

$$B = \begin{pmatrix} 1 & 0 \\ 0 & 1 \\ 0 & 0 \end{pmatrix},\quad v_0 = \begin{pmatrix} v_{01} \\ v_{02} \end{pmatrix},\quad x = \begin{pmatrix} x_1 \\ x_2 \\ x_3 \end{pmatrix}.$$

Suppose that the output $y(t)$ is equal to $x_3(t)$, that is, equal to $x_3(t)$. Then,

$$y(t) = Cx(t),\quad \text{where}\quad C = (0\ \ 0\ \ 1). \tag{4.63}$$

We also assume that the following interval initial conditions are given:

$$x(t_0) = x_0 \in [\underline{x}_0, \overline{x}_0]. \tag{4.64}$$

In order to use Theorem 4.3 for solving the formulated problem, let us find out under what conditions assumptions D_3–D_6 are satisfied. Since $G(0) = 0$, then assumption D_3 is satisfied.

Consider the matrix $A(t) - BK_0C$ from assumption D_4. In our case, K_0 has a dimension of 2×1, so the matrix K_0 can be represented as $K_0 = \begin{pmatrix} k_1 \\ k_2 \end{pmatrix}$.

Denote by a_{ij} $(i, j = 1, 2, 3)$ the elements of the matrix A, while $a_{33} = 0$. Then, $A - BK_0C$ is equal to

$$A - BK_0C = \begin{pmatrix} a_{11} & a_{12} & a_{13} - k_1 \\ a_{21} & a_{22} & a_{23} - k_2 \\ a_{31} & a_{32} & 0 \end{pmatrix}.$$

Since the matrix $A(t) - BK_0C$ is constant, the fundamental matrix of the system from assumption D_4 is equal to the exponential of the difference of the matrices $A - BK_0C$; therefore, the condition (4.51) will be true if and only if

$$\max_i \operatorname{Re} \lambda_i (A - BK_0C) < 0. \tag{4.65}$$

As is known, the fulfilment of the conditions (4.65) can be ensured by the fulfilment of the *Routh–Hurwitz conditions* for the characteristic polynomial of the matrix $A - BK_0C$. For the characteristic polynomial of the third order

$$f(z) = z^3 + a_1 z^2 + a_2 z + a_3,$$

these conditions are

$$a_1 > 0, \quad a_1 a_2 - a_3 > 0, \quad a_3 > 0. \tag{4.66}$$

Having calculated the *characteristic polynomial* for the matrix $A - BK_0C$, we can rewrite the conditions (4.66) as

$$\begin{cases} -a_{11} - a_{22} > 0, \\ (-a_{11} - a_{22})(a_{11}a_{22} - a_{31}a_{13} + a_{31}k_1 - a_{32}a_{23} + a_{32}k_2 - a_{12}a_{21}) \\ + a_{12}a_{31}a_{23} - a_{12}a_{31}k_2 + a_{21}a_{32}a_{13} - a_{21}a_{32}k_1 - a_{31}a_{22}a_{13} \\ + a_{31}a_{22}k_1 - a_{32}a_{11}a_{23} + a_{32}a_{11}k_2 > 0, \\ - a_{12}a_{31}a_{23} + a_{12}a_{31}k_2 - a_{21}a_{32}a_{13} + a_{21}a_{32}k_1 + a_{31}a_{22}a_{13} \\ - a_{31}a_{22}k_1 + a_{32}a_{11}a_{23} - a_{32}a_{11}k_2 > 0. \end{cases} \tag{4.67}$$

The first inequality (4.67) holds because $a_{11} = -\frac{R_s}{L_d} < 0$, $a_{22} = -\frac{R_s}{L_q} < 0$. Then, the system (4.67) can be rewritten as

$$MK_0 > L, \qquad (4.68)$$

where

$$
M = \begin{pmatrix} -a_{11}a_{31} - a_{22}a_{31} & -a_{22}a_{32} - a_{12}a_{31} \\ -a_{21}a_{32} + a_{31}a_{22} & -a_{11}a_{31} + a_{32}a_{11} \end{pmatrix},
$$

$$
L = (L_{11}, L_{21})^T,
$$

$$
L_{11} = a_{11}^2 a_{22} - a_{11}a_{31}a_{13} - a_{11}a_{12}a_{21} + a_{22}^2 a_{11} - a_{22}a_{32}a_{23}
$$

$$
- a_{22}a_{12}a_{21} - a_{12}a_{31}a_{23} - a_{21}a_{32}a_{13},
$$

$$
L_{21} = a_{12}a_{31}a_{23} + a_{21}a_{32}a_{13} - a_{31}a_{22}a_{13} - a_{32}a_{11}a_{23},
$$

$$(4.69)$$

and the inequality is considered element by element.

Thus, if K_0 satisfies the inequality (4.68), then condition D_4 is satisfied.

Consider assumptions D_5 and D_6. Let us estimate the function G in the norm

$$
\|G(x(t))\|
$$

$$
\leq \left| \frac{3n_p i_{q0}}{2J_0}(L_d - L_q)x_1 + \left(\frac{3n_p i_{d0}}{2J_0}(L_d - L_q) + \frac{3n_p \Phi}{2} \right) x_2 \right.
$$

$$
\left. + \frac{3n_p}{2J_0}(L_d - L_q)x_1 x_2 \right| \leq \left| \left(\frac{3n_p}{2J_0}(L_d - L_q)(i_{q0} + x_2) \right) x_1 \right.
$$

$$
\left. + \left(\frac{3n_p i_{d0}}{2J_0}(L_d - L_q) + \frac{3n_p \Phi}{2} \right) x_2 \right| \leq \gamma \|x\|, \qquad (4.70)
$$

where

$$
\gamma = \sup_{x \in D} \sqrt{ \left(\frac{3n_p}{2J_0}(L_d - L_q)(i_{q0} + x_2) \right)^2 + \left(\frac{3n_p i_{d0}}{2J_0}(L_d - L_q) + \frac{3n_p \Phi}{2} \right)^2 },
$$

and D is some region in $\mathbb{R}^3$ containing 0. Similarly, we have an estimate for $Y(x(t))$:

$$
\|Y(x(t))\| \leq \|r\| \|x\|^2, \qquad (4.71)
$$

where $r = \left(\frac{n_p L_q}{L_d}, -\frac{n_p L_d}{L_q}, \frac{3n_p}{2J_0}(L_d - L_q) \right)^T$. It follows from (4.70) and (4.71) that conditions D_5 and D_6 are satisfied for any values of the system parameters. Thus, we can formulate the stability conditions for the equilibrium state (4.60) of the system (4.62)–(4.63) under the interval initial conditions (4.64).

If the inequality (4.68) *is solvable with respect to* K_0, *then the controls*

$$u = -Ky, \quad v_0 = -K_0 y,$$

where $K \in \mathbb{R}$ *and* K_0 *is the solution of* (4.68), *stabilise the motion of the system* (4.62) *and* (4.63) *under the interval initial conditions* (4.64).

4.7   Finite Interval Stability of Affine Systems

In this section, sufficient conditions for the stability of motion on a *finite interval* of an uncertain *affine system* are obtained under interval initial conditions. These conditions are obtained on the basis of new estimates for the norms of solutions of the corresponding systems of equations of perturbed motion. In this case, a pseudo-linear representation of the nonlinear integral inequalities used in the estimates under consideration is applied.

4.7.1   Notation and definitions

Let $\mathbb{R}^n$ be an n-dimensional Euclidean space and $J = [t_0, t_0 + T]$ be any finite interval on $\mathbb{R}_+$, $T = const > 0$. For any sets A and B in $\mathbb{R}^n$, the expression $A \times B$ denotes the Cartesian product and $\emptyset$ is the empty set.

The system of equations of perturbed motion of some mechanical system is considered in the form

$$\frac{dx}{dt} = f(t, x), \tag{4.72}$$

$$x(t_0) = x_0 \in [\underline{x}_0, \overline{x}_0] \tag{4.73}$$

under the interval initial conditions x_0. Here, $x \in \mathbb{R}^n$ and $f : \mathbb{R}_+ \times \mathbb{R}^n \to \mathbb{R}^n$. The bundle of solution of the system (4.72) is denoted by $x(t, t_0, x_0)$, with the initial conditions (4.73). For an interval vector $x = (x_1, \ldots, x_n)^T \in \mathbb{R}^n$, the norm of x is denoted as follows:

$$\|x\|_I = \max\{|x_1|, \ldots, |x_n|\}, \tag{4.74}$$

where $|x_i| = \max\{|\underline{x}_i|, |\overline{x}_i|\}$ for each value $i = 1, 2, \ldots, n$. The norm of the interval matrix is defined similarly:

$$\|A\|_I = \max_i \sum_{j=1}^{n} |A_{ij}|, \quad i, j = 1, 2, \ldots, n. \tag{4.75}$$

Along with the *interval norms* (4.74) and (4.75), we consider the usual Euclidean norm $\|\cdot\|$ of the vector x and the matrix A. For given values of λ

and A in the space $\mathbb{R}^n$, we consider the domains $S_\lambda = \{x \in \mathbb{R}^n : \|x\| < \lambda\}$ and $S_A = \{x \in \mathbb{R}^n : \|x\| < A\}$.

Next, we need the following definitions.

Definition 4.3. The system (4.72) is (λ, A, J)-*stable* if, for given values λ, A, J, and $0 < \lambda < A$, from the condition $(x_0 \in [\underline{x}_0, \overline{x}_0]) \cap (\|x_0\|_I < \lambda)$, it follows that $\|x(t, t_0, x_0)\|_I < A$ for all $t \in J$, $J = [t_0, t_0 + T]$.

Definition 4.4. The system (4.72) is *uniformly in* t_0 (λ, A, J)-*stable* if the conditions of Definition 4.3 are satisfied for any $t_0 \in \mathbb{R}_+$.

Everywhere in the following, we consider the interval norm of a vector in the form (4.74).

For a system of differential equations (4.72), the answer to the following question is of interest: under what restrictions on the parameters of the system (4.72), the bundle of its solutions $x(t, t_0, x_\beta)$ at $x_\beta \in [\underline{x}_0, \overline{x}_0]$, $x_\beta = \underline{x}_0 \beta + (1 - \beta)\overline{x}_0$, $\beta \in [0, 1]$ has a certain type of (λ, A, J)-stability. It is assumed that, for any fixed value of x_β from the interval $[\underline{x}_0, \overline{x}_0]$, the solution of the system (4.72) exists for all $t \in J$.

4.7.2 Stability of a quasi-linear system

Let the vector function $f(t, x)$ in the system (4.72) have the form

$$f(t, x) = A(t)x + g_1(t, x), \tag{4.76}$$

where $A(t)$ is an $n \times n$ matrix with continuous entries for any $t \in J$ and $g_1(t, x) \in C(J \times \mathbb{R}^n, \mathbb{R}^n)$ is sufficiently smooth, so that a solution to the system (4.72) with (4.76) exists on J for any $x_0 \in [\underline{x}_0, \overline{x}_0]$.

Let us show that the following assertion holds.

Theorem 4.4. *Suppose that:*

(1) *for all $t \in J$, there exists a non-negative integrable function $b(t)$ such that*

$$\|A(t)\| \le b(t) \quad \text{for all} \quad t \in J;$$

(2) *for all $t \in J$, there exists a non-negative integrable function $c(t)$ such that*

$$\|g_1(t, x)\| \le c(t)\|x\|^\alpha, \quad \alpha > 1,$$

for all $(t, x) \in J \times \mathbb{R}^n$;

(3) *for all $t \in J$, the inequality*

$$(\alpha - 1)\|x_0\|_I^{\alpha-1} \int_{t_0}^{t} c(s) \exp\left[(\alpha - 1) \int_{t_0}^{s} b(\tau)d\tau\right] ds < 1;$$

(4) *given estimates for the quantities λ and A, for all $t \in J$, the following inequality holds:*

$$\frac{\exp\left(\int_{t_0}^{t} b(s)ds\right)}{\left[1 - (\alpha - 1)\lambda^{\alpha-1} \int_{t_0}^{t} c(s) \exp\left[(\alpha - 1) \int_{t_0}^{s} b(\tau)d\tau\right] ds\right]^{\frac{1}{\alpha-1}}} < \frac{A}{\lambda}.$$

$$(4.77)$$

Then, the system (4.72) with the right-hand side of (4.76) is (λ, A, J)-stable.

Proof. From the system of equations

$$\frac{dx}{dt} = A(t)x + g_1(t, x), \tag{4.78}$$

$$x(t_0) = x_0 \in [\underline{x}_0, \overline{x}_0], \tag{4.79}$$

we get the relation

$$x(t, t_0, x_0) = x_0 + \int_{t_0}^{t} (A(s)x(s) + g_1(s, x(s)))ds \tag{4.80}$$

for any $x_0 \in [\underline{x}_0, \overline{x}_0]$. From (4.80), we get the estimate

$$\|x(t, t_0, x_0)\| \leq \|x_0\|_I + \int_{t_0}^{t} (\|A(s)\|\|x(s)\| + \|g_1(s, x(s))\|)ds$$

for all $t \in J$. Taking into account conditions (1) and (2) of Theorem 4.4, we obtain

$$\|x(t, t_0, x_0)\| \leq \|x_0\|_I + \int_{t_0}^{t} (b(s)\|x(s)\| + c(s)\|x(s)\|^\alpha)ds \tag{4.81}$$

for all $t \in J$. Let $\|x(t, t_0, x_0)\| = \psi(t)$ for all $t \in J$. From the inequality (4.81), it follows that

$$\psi(t) \leq \|x_0\|_I + \int_{t_0}^{t} (b(s) + c(s)\psi^{\alpha-1}(s))\psi(s)\, ds. \tag{4.82}$$

$\square$

Applying Lemma 1.2 to the inequality (4.82), we obtain the estimate

$$\psi(t) \leq \|x_0\|_I \exp\left(\int_{t_0}^t (b(s) + c(s)\psi^{\alpha-1}(s))\,ds\right). \tag{4.83}$$

As in the proof of Lemma 1.6, it is easy to obtain the estimate

$$\exp\left[(\alpha-1)\int_{t_0}^t c(s)\psi^{\alpha-1}(s)\,ds\right]$$

$$\leq \left[1 - (\alpha-1)\|x_0\|_I^{\alpha-1}\int_{t_0}^t c(s)\exp\left((\alpha-1)\int_{t_0}^s b(\tau)\,d\tau\right)ds\right]^{-1} \tag{4.84}$$

for all $t \geq s \geq t_0$ and $t \in J$. Moreover, by virtue of condition (3) of Theorem 4.4, the expression on the right-hand side of the inequality (4.84) exists for all $t \in J$. Taking into account the inequality (4.84), we obtain the estimate

$$\psi^{\alpha-1}(t) \leq \frac{\|x_0\|_I^{\alpha-1} \exp\left((\alpha-1)\int_{t_0}^t b(s)\,ds\right)}{1 - (\alpha-1)\|x_0\|_I^{\alpha-1}\int_{t_0}^t c(s)\exp\left((\alpha-1)\int_{t_0}^s b(\tau)\,d\tau\right)ds}. \tag{4.85}$$

The inequality (4.85) implies an estimate for the norm of solutions to the system (4.76):

$$\|x(t,t_0,x_0)\|_I$$

$$\leq \frac{\|x_0\|_I \exp\left(\int_{t_0}^t b(s)\,ds\right)}{\left[1 - (\alpha-1)\|x_0\|_I^{\alpha-1}\int_{t_0}^t c(s)\exp\left((\alpha-1)\int_{t_0}^s b(\tau)\,d\tau\right)ds\right]^{\frac{1}{\alpha-1}}}. \tag{4.86}$$

Using the bound (4.86), we now show that the system (4.76) is (λ, A, J)-stable. Assume that for a given $\lambda > 0$, the initial values satisfy the condition $x_0 \in [\underline{x}_0, \overline{x}_0]$ and $\|x_0\|_I < \lambda$. Then, from (4.86), we obtain an estimate of the norms of solutions of the norms of bunch solutions under given initial conditions:

$$\|x(t,t_0,x_0)\|$$

$$\leq \frac{\lambda \exp\left(\int_{t_0}^t b(s)\,ds\right)}{\left[1 - (\alpha-1)\lambda^{\alpha-1}\int_{t_0}^t c(s)\exp\left((\alpha-1)\int_{t_0}^s b(\tau)\,d\tau\right)ds\right]^{\frac{1}{\alpha-1}}}. \tag{4.87}$$

for all $t \in J$. Using condition (4) of Theorem 4.4, from (4.87), we obtain

$$\|x(t, t_0, x_0)\| < \lambda \cdot \frac{A}{\lambda} = A$$

for all $t \in J$, which means (λ, A, J)-stability of the system (4.76). Theorem 4.4 is proved.

Remark 4.1. If condition (3) of Theorem 4.4 and the inequality (4.77) hold for all $t \in [t_0, \infty)$, $t_0 \in \mathbb{R}_+$, then the system (4.76) is (λ, A, ∞)-*stable*, i.e., practically stable (see [67, 70] and the bibliography therein).

4.7.3 Stability of essentially nonlinear systems

The following system of equations of perturbed motion is considered:

$$\frac{dx}{dt} = g_1(t, x) + g_2(t, x), \tag{4.88}$$

$$x(t_0) = x_0 \in [\underline{x}_0, \overline{x}_0], \tag{4.89}$$

where $g_1(t, x), g_2(t, x) \in C(J \times \mathbb{R}^n, \mathbb{R}^n)$ are sufficiently smooth so that the solution to the problem (4.88)–(4.89) exists on J for any $x_0 \in [\underline{x}_0, \overline{x}_0]$. Regarding the system (4.88), suppose the following:

D_{11}. There exist integrable non-negative functions $h_1(t)$ and $h_2(t)$ and constants $p > 1$ and $q \geq 1$ such that for all $(t, x) \in J \times S_A$,

$$\|g_1(t, x)\| \leq h_1(t)\|x\|^p,$$

$$\|g_2(t, x)\| \leq h_2(t)\|x\|^q.$$

Under the condition D_{11}, from the relation

$$x(t, t_0, x_0) = x_0 + \int_{t_0}^t (g_1(s, x(s)) + g_2(s, x(s))) \, ds, \tag{4.90}$$

we get the inequality

$$\|x(t, t_0, x_0)\| \leq \|x_0\|_I + \int_{t_0}^t (h_1(s)\|x(s)\|^p + h_2(s)\|x(s)\|^q) \, ds \tag{4.91}$$

for all $t \in J$.

If the condition D_{11} is satisfied for the system (4.88), the following assertion takes place.

Theorem 4.5. *Let the inequality* (4.91) *hold, and for x_0 such that $x_0 \in [\underline{x}_0, \overline{x}_0]$ and $\|x_0\| < \lambda$, the following estimate is correct:*

$$\frac{\lambda}{\left(1 - (p + q - 2)\left(\lambda^{p-1}\int_{t_0}^{t} h_1(s)ds + \lambda^{q-1}\int_{t_0}^{t} h_2(s)ds\right)\right)^{\frac{1}{p+q-2}}} < A \tag{4.92}$$

for all $t \in J$. Then, the solution $x(t, t_0, x_0)$ of the system (4.88) *is (λ, A, J)-stable.*

Theorem 4.6. *If the estimate* (4.91) *holds for any $t_0 \in \mathbb{R}_+$, then the solution $x(t, t_0, x_0)$ of the system* (4.88) *is uniformly in t_0 (λ, A, J)-stable.*

Proofs of Theorems 4.5 and 4.6 follow from an estimate of the norm of solution to the system (4.88) and the inequality (4.92).

4.7.4 Stabilisation of an interval affine system

Consider a nonlinear *non-autonomous affine system* with many controls of the form (cf. [72, 125])

$$\frac{dx}{dt} = (A_0 + \Delta A)x(t) + Y(t, x) + \sum_{i=1}^{m} G_i(t, x(t))u_i(t)$$

$$+ (B_0 + \Delta B)u_0(t), \tag{4.93}$$

$$y(t) = (C_0 + \Delta C)x(t), \tag{4.94}$$

$$x(t_0) = x_0 \in [\underline{x}_0, \overline{x}_0]. \tag{4.95}$$

Here, the vector function $Y(t, x)$ is such that the set of solutions of the system (4.93) under the interval initial conditions (4.94) exists for all $t \in J$, $x \in S_A$, A_0 is a nominal $n \times n$ matrix, $G_i(t, x)$ is an $n \times m$-matrix; the control vectors $u_i(t) \in \mathbb{R}^m$ for all $i = 1, 2, \ldots, m$; B_0 is an $n \times m$ matrix; control $u_0(t) \in \mathbb{R}^m$; and C_0 is a constant $n \times n$ matrix; ΔA and ΔB are some matrices of appropriate dimensions, taking into account the inaccuracy of the system (4.93); and x_0 is the interval vector of initial states of the system (4.93). We make the following assumptions about the system (4.93):

D_{12}. Functions $G_i(t, 0) = 0$, $i = 1, 2, \ldots, m$, for all $t \geq 0$.
D_{13}. There exists a constant $n \times m$ matrix K_0 such that for the system

$$dy/dt = (A_0 - B_0K_0C_0)y,$$

the fundamental matrix $\Phi(t)$ satisfies the estimate

$$\|\Phi(t)\Phi^{-1}(s)\| \le Me^{-\alpha(t-s)} \tag{4.96}$$

for $t \ge s \ge t_0$, where M and α are some positive constants.

D_{14}. There are constants $\gamma_i > 0$ and $q \ge 1$ such that

$$\|G_i(t, x)\| \le \gamma_i \|x\|^q \tag{4.97}$$

for all $i = 1, 2, \ldots, m$.

D_{15}. There are constants $\varkappa > 0$ and $p > 0$ such that

$$\|Y(t, x)\| \le \varkappa \|x\|^{p+1} \tag{4.98}$$

in the range $(t, x) \in \mathbb{R}_+ \times D$.

In what follows, we need the notations

$$\gamma = \sum_{i=1}^{m} \gamma_i,$$

$$D = \Delta A - \Delta B K_0 C_0 - \Delta B K_0 \Delta C - B_0 K_0 \Delta C,$$

$$h = M\|D\| - \alpha,$$

$$\Theta_{11} = M^{p+1} e^{-pht_0} \chi(p+q),$$

$$\Theta_{21} = M^{q+1} e^{-qht_0} \gamma(p+q) \sum_{i=1}^{m} \|K_i(C_0 + \Delta C)\|.$$

We also define the functions $\sigma(u)$ and $F(u, t)$ as follows:

$$\sigma(u) = \sup\left\{ t : \Theta_{11} u^p \int_{t_0}^{t} e^{phs} ds + \Theta_{21} u^q \int_{t_0}^{t} e^{qhs} ds < 1, \, t \ge t_0 \right\},$$

$$u > 0;$$

$$F(u, t) = \frac{Mu e^{h(t-t_0)}}{\left(1 - \Theta_{11} u^p \int_{t_0}^{t} e^{phs} ds - \Theta_{21} u^q \int_{t_0}^{t} e^{qhs} ds \right)^{\frac{1}{p+q}}},$$

$$(u, t) \in \mathcal{N},$$

where

$$\mathcal{N} = \left\{ (u, t) : \Theta_{11} u^p \int_{t_0}^{t} e^{phs} ds + \Theta_{21} u^q \int_{t_0}^{t} e^{qhs} ds < 1, \, t \ge t_0, \, u > 0 \right\}.$$

Theorem 4.7. *Let the conditions of assumptions A_2–A_5 be fulfilled. Then, for the controls*

$$u_i(t) = -K_i y(t), \quad i = 1, 2, \ldots, m, \quad u_0(t) = -K_0 y(t), \tag{4.99}$$

the matrices $K_i (i = 1, 2, \ldots, m)$ and K_0 are correlated with the parameters of the system (4.93)–(4.94) by the relations

$$\sigma(\lambda) > t_0 + T,$$

$$\begin{cases} F(\lambda, t_0 + T) < A, \\ h \geq 0; \end{cases} \quad \vee \quad \begin{cases} F(\lambda, t_0 + T) < A, \\ F(\lambda, t_0) < A, \\ h < 0. \end{cases}$$

Proof. Let $m < n$. For the stabilisation of the set of trajectories of the system (4.93) under the initial conditions (4.95), the controls (4.99) are used. From (4.93), under the controls (4.98), it follows that

$$\frac{dx}{dt} = (A_0 + \Delta A)x(t) + Y(t, x(t))$$

$$+ \sum_{i=1}^{m} G_i(t, x(t))(-K_i)(C_0 + \Delta C)x(t)$$

$$+ (B_0 + \Delta B)(-K_0)(C_0 + \Delta C)x(t)$$

$$= (A_0 - B_0 K_0 C_0)x(t) + (\Delta A - \Delta B K_0 C_0$$

$$- \Delta B K_0 \Delta C - B_0 K_0 \Delta C)x(t) + Y(t, x(t))$$

$$- \sum_{i=1}^{m} G_i(t, x(t))K_i(C_0 - \Delta C)x(t).$$

Applying the constant variation formula to the resulting system, we obtain

$$\square$$

$$x(t, t_0, x_0) = \Phi(t)\Phi^{-1}(t_0)x_0$$

$$+ \int_{t_0}^{t} \Phi(t)\Phi^{-1}(s)\left(Dx(s) + Y(s, x(s)) \right.$$

$$\left. - \sum_{i=1}^{m} G_i(s, x(s))K_i(C_0 + \Delta C)x(s) \right) ds, \tag{4.100}$$

where $x_0 \in [\underline{x}_0, \overline{x}_0]$. In view of the estimates (4.96)–(4.98), it follows from the relation (4.100) that

$$\|x(t, t_0, x_0)\| \leq \|x_0\|_I M e^{-\alpha(t-t_0)} + e^{-\alpha t}$$

$$\times \int_{t_0}^{t} \left(M\|D\| e^{\alpha s}\|x(s)\| + M\chi e^{\alpha s}\|x(s)\|^{p+1} \right.$$

$$\left. + \sum_{i=1}^{m} M\gamma\|K_i(C_0 + \Delta C)\| e^{\alpha s}\|x(s)\|^{q+1} \right) ds. \qquad (4.101)$$

Next, we transform the inequality (4.101) to the form

$$\|x(t, t_0, x_0)\| e^{\alpha t} \leq \|x_0\|_I M e^{\alpha t_0}$$

$$+ \int_{t_0}^{t} \left(M\|D\| \|x(s)\| e^{\alpha s} + M\chi e^{-p\alpha s}(\|x(s)\| e^{\alpha s})^{p+1} \right.$$

$$\left. + M\gamma \sum_{i=1}^{m} \|K_i(C_0 + \Delta C)\| e^{-q\alpha s}(\|x(s)\| e^{\alpha s})^{q+1} \right) ds. \qquad (4.102)$$

Denote $\|x(t, t_0, x_0)\| e^{\alpha t} = \psi(t)$. Then, (4.102) becomes

$$\psi(t) \leq \|x_0\| M e^{\alpha t_0} + \int_{t_0}^{t} \left(M\|D\| + M\chi e^{-p\alpha s}\psi^p(s) \right.$$

$$\left. + M\gamma \sum_{i=1}^{m} \|K_i(C_0 + \Delta C)\| e^{-q\alpha s}\psi^q(s) \right) \psi(s) ds. \qquad (4.103)$$

Applying Lemma 1.2 to the inequality (4.103) thus obtained, we get the estimate

$$\psi(t) \leq \|x_0\| M e^{\alpha t_0} \cdot \exp\left\{ \int_{t_0}^{t} \left(M\|D\| + M\chi e^{-p\alpha s}\psi^p(s) \right. \right.$$

$$\left. \left. + M\gamma \sum_{i=1}^{m} \|K_i(C_0 + \Delta C)\| e^{-q\alpha s}\psi^q(s) \right) ds \right\},$$

or

$$\psi(t) \le \|x_0\| M \cdot \exp\left\{ \alpha t_0 + M\|D\|(t - t_0) \right\}$$

$$\times \exp\left\{ \int_{t_0}^{t} \left(M\chi e^{-pas} \psi^p(s) \right. \right.$$

$$\left. \left. + M\gamma \sum_{i=1}^{m} \|K_i(C_0 + \Delta C)\| e^{-qas} \psi^q(s) \right) ds \right\}. \tag{4.104}$$

Denote $\psi(t)e^{-M\|D\|t} = \varphi(t)$. Then, (4.104) becomes

$$\varphi(t) \le \|x_0\| M \cdot \exp\left\{ (\alpha - M\|D\|)t_0 \right\}$$

$$\times \exp\left\{ \int_{t_0}^{t} \left(M\chi e^{-ps(\alpha - M\|D\|)} \varphi^p(s) \right. \right.$$

$$\left. \left. + M\gamma \sum_{i=1}^{m} \|K_i(C_0 + \Delta C)\| e^{-qs(\alpha - M\|D\|)} \varphi^q(s) \right) ds \right\}, \tag{4.105}$$

or

$$\varphi(t) \le \|x_0\| M e^{-ht_0}$$

$$\times \exp\left\{ \int_{t_0}^{t} \left(M\chi e^{phs} \varphi^p(s) + M\gamma \sum_{i=1}^{m} \|K_i(C_0 + \Delta C)\| e^{qhs} \varphi^q(s) \right) ds \right\}.$$

We introduce further the notation

$$\begin{cases} k_1(t) = M\chi e^{pht}, \\ k_2(t) = M\gamma \sum_{i=1}^{m} \|K_i(C_0 + \Delta C)\| e^{qht}, \\ \varphi_0 = \|x_0\| M e^{-ht_0}. \end{cases} \tag{4.106}$$

Then, (4.105) becomes

$$\varphi(t) \le \varphi_0 \cdot \exp\left\{ \int_{t_0}^{t} (k_1(s)\varphi^p(s) + k_2(s)\varphi^q(s)) ds \right\}. \tag{4.107}$$

From the estimate (4.107), it is easy to obtain the following inequality:

$$\varphi(t) \le \frac{\varphi_0}{\left[1 - r \left(\varphi_0^p \int_{t_0}^t k_1(s)ds + \varphi_0^q \int_{t_0}^t k_2(s)ds \right) \right]^{\frac{1}{r}}},$$

where $r = p + q - 2$.

Returning to the notation (4.105), we get

$$\varphi(t)$$

$$\le \frac{\varphi_0}{\left[1 - r \left(\varphi_0^p \int_{t_0}^t M\chi e^{phs}ds + \varphi_0^q \int_{t_0}^t M\gamma \sum_{i=1}^m \|K_i(C_0 + \Delta C)\| e^{qhs}ds \right) \right]^{\frac{1}{r}}}.$$

In view of the notation introduced earlier, the last inequality implies an estimate for the norm of solutions of the system (4.93) in the form

$$\|x(t,t_0,x_0)\| = \psi(t)e^{-\alpha t} = \varphi(t)e^{M\|D\|t}e^{-\alpha t} = \varphi(t)e^{ht} \le F(\|x_0\|_I, t). \tag{4.108}$$

The estimate (4.108) is valid for all $(\|x_0\|_I, t) \in \mathcal{N}$. Using this estimate, we show that the solution $x = 0$ of the system (4.93) is (λ, A, J)-stable.

Assume that, given $\lambda > 0$, the initial values satisfy the conditions $x_0 \in [\underline{x}_0, \overline{x}_0]$ and $\|x_0\| < \lambda$. Then, the estimate (4.108) holds for any $t \in J$. Indeed, consider the function

$$R(u,t) = \Theta_{11}u^p \int_{t_0}^t e^{phs}ds + \Theta_{21}u^q \int_{t_0}^t e^{qhs}ds,$$

which is defined for all $u > 0$ and $t \ge t_0$. The function $R(u,t)$ increases in t on the interval $[t_0, +\infty)$ for any fixed value of $u > 0$, so the equation $R(u,t) = 1$ with respect to the variable t has at most one solution. Then, the inequality $R(u,t) < 1$ for a given u holds for all $t \in [t_0, +\infty)$ if the equation has no solution and for all $t \in [t_0, \sigma(u))$ if this equation has a solution. In addition, the choice of initial conditions implies the inclusion

$$\left\{ t : \Theta_{11}\lambda^p \int_{t_0}^t e^{phs}ds + \Theta_{21}\lambda^q \int_{t_0}^t e^{qhs}ds < 1, t \ge t_0 \right\}$$

$$\subset \left\{ t : \Theta_{11}\|x_0\|_I^p \int_{t_0}^t e^{phs}ds + \Theta_{21}\|x_0\|_I^q \int_{t_0}^t e^{qhs}ds < 1, t \ge t_0 \right\},$$

which yields the inequality

$$\sigma(\lambda) \le \sigma(\|x_0\|_I).$$

Therefore, the inequality $R(\|x_0\|_I, t) < 1$ holds for any x_0 such that $x_0 \in [\underline{x}_0, \overline{x}_0]$ and $\|x_0\|_I < \lambda$ and for all $t \in [t_0, \sigma(\lambda))$. Due to the inequality $\sigma(\lambda) > t_0 + T$, the inclusion $t \in [t_0, \sigma(\lambda))$ implies the inclusion $t \in [t_0, t_0 + T]$, i.e., $t \in J$.

Thus, (4.108) is evaluated for any $t \in J$ as soon as $x_0 \in [\underline{x}_0, \overline{x}_0]$ and $\|x_0\|_I < \lambda$.

Since the function $F(u, t)$ is monotonically increasing in the variable $u > 0$, for any fixed value of $t \in J$, (4.108) also implies the estimate

$$\|x(t, t_0, x_0)\| \le F(\|x_0\|_I, t) < F(\lambda, t), \tag{4.109}$$

for all $t \in J$.

Let us examine the behaviour of $F(\lambda, \cdot)$. Compute $\frac{\partial F}{\partial t}$ as

$$\frac{\partial F}{\partial t} = \frac{\partial}{\partial t} \left(\frac{M\lambda e^{h(t-t_0)}}{\left(1 - \Theta_{11}\lambda^p \int_{t_0}^t e^{phs}ds - \Theta_{21}\lambda^q \int_{t_0}^t e^{qhs}ds \right)^{\frac{1}{p+q}}} \right)$$

$$= \frac{\partial}{\partial t} \left(\frac{M\lambda e^{h(t-t_0)}}{(1 - R(\lambda, t))^{\frac{1}{p+q}}} \right)$$

$$= \frac{M\lambda e^{h(t-t_0)} \left(h(p+q)\left(1 - R(\lambda, t)\right) + \frac{\partial R}{\partial t} \right)}{(p+q)\left(1 - R(\lambda, t)\right)^{1+\frac{1}{p+q}}}$$

$$= \frac{M\lambda e^{h(t-t_0)} \left(h(p+q)\left(1 - R(\lambda, t)\right) + \Theta_{11}\lambda^p e^{pht} + \Theta_{21}\lambda^q e^{qht} \right)}{(p+q)\left(1 - R(\lambda, t)\right)^{1+\frac{1}{p+q}}}.$$

(1) Suppose that $h \ge 0$. Then, $\frac{\partial F(\lambda, t)}{\partial t} > 0$, for all $t \in J$, i.e., $F(\lambda, \cdot)$ is increasing on the interval J. Then, $F(\lambda, t) \le F(\lambda, t_0 + T)$ for all $t \in J = [t_0, t_0 + T]$. Therefore, if $h \ge 0$, then for given $\lambda > 0$ and $A > 0$, from the inequality $F(\lambda, t_0 + T) < A$ of condition ??Q_1 of Theorem 4.7, it follows that the estimate for the norm of solutions

$$\|x(t, t_0, x_0)\| < A$$

holds for all $t \in J$, that is, the solution $x = 0$ of the system (4.93) is (λ, A, J)-stable under the interval initial conditions (4.95).

(2) Suppose now that $h < 0$. Then, it is easy to see that $\frac{\partial F(\lambda, t)}{\partial t}$ on the interval $t \in J$ or takes one sign, that is, $F(\lambda, \cdot)$ is increasing on the interval J, or $\frac{\partial F(\lambda, t)}{\partial t}$ is equal to zero at the only point $t^* \in J$. Moreover, for sufficiently small $\delta > 0$, $\frac{\partial F(\lambda, t^*-\delta)}{\partial t} \frac{\partial F(\lambda, t^*+\delta)}{\partial t} < 0$, that is, t^* is a

local minimum point of $F(\lambda, \cdot)$ on the interval J. Therefore, for a given $\lambda > 0$, we have

$$\max_{t \in [t_0, t_0 + T]} F(\lambda, t) = \max \{F(\lambda, t_0), F(\lambda, t_0 + T)\}.$$

If we now use the inequalities $F(\lambda, t_0 + T) < A$ and $F(\lambda, t_0) < A$ from condition Q_2 of Theorem 4.7, we obtain the following estimate for the norm of bunch solutions under given initial conditions:

$$\|x(t, t_0, x_0)\| < F(\lambda, t) \leq \max \{F(\lambda, t_0), F(\lambda, t_0 + T)\} < A,$$

that is, the solution $x = 0$ of the system (4.93) is (λ, A, J)-stable under the interval initial conditions (4.95).

Theorem 4.7 is proved.

Suppose that under the conditions of Theorem 4.7, the functions $G_i(t, x)$ satisfy the estimate obtained by modifying the estimate (4.97):

$$\|G_i(t, x)\| \leq \gamma_{i1} \|x\|^{q_1} + \gamma_{i2} \|x\|^{q_2} \tag{4.110}$$

for all $i = 1, 2, \ldots, m$, where $\gamma_{i1}, \gamma_{i2} > 0$ $(i = 1, 2, \ldots, m)$, $q_1, q_2 \geq 1$.

Then, a similar theorem holds.

Theorem 4.8. *Let the conditions of assumptions D_{12}–D_{15} be satisfied when the inequality (4.96) is replaced by the inequality (4.110). Then, under the controls (4.99), the matrices K_i $(i = 1, 2, \ldots, m)$ and K_0 are related to the parameters of the system (4.93) by the relations*

$$\widetilde{\sigma}(\lambda) > t_0 + T, \tag{4.111}$$

$$\begin{cases} \widetilde{F}(\lambda, t_0 + T) < A, \\ h \geq 0; \end{cases} \quad \vee \quad \begin{cases} \widetilde{F}(\lambda, t_0 + T) < A, \\ \widetilde{F}(\lambda, t_0) < A, \\ h < 0, \end{cases} \tag{4.112}$$

where

$$\widetilde{\sigma}(u) = \sup \Big\{ t : \widetilde{\Theta}_{11} u^p \int_{t_0}^t e^{phs} ds$$

$$+ \widetilde{\Theta}_{21} u^{q_1} \int_{t_0}^t e^{q_1 hs} ds + \widetilde{\Theta}_{22} u^{q_2} \int_{t_0}^t e^{q_2 hs} ds < 1, \ t \geq t_0 \Big\}, \quad u > 0,$$

$\widetilde{F}(u,t)$

$$= \frac{Mue^{h(t-t_0)}}{\left(1 - \widetilde{\Theta}_{11}u^p \displaystyle\int_{t_0}^t e^{phs}ds - \widetilde{\Theta}_{21}u^{q_1}\displaystyle\int_{t_0}^t e^{q_1 hs}ds - \widetilde{\Theta}_{22}u^{q_2}\displaystyle\int_{t_0}^t e^{q_2 hs}ds\right)^{\frac{1}{p+q}}},$$

$(u,t) \in \widetilde{\mathcal{N}},$

$$\widetilde{\mathcal{N}} = \Big\{(u,t) : \widetilde{\Theta}_{11}u^p \int_{t_0}^t e^{phs}ds$$

$$+ \widetilde{\Theta}_{21}u^{q_1}\int_{t_0}^t e^{q_1 hs}ds + \widetilde{\Theta}_{22}u^{q_2}\int_{t_0}^t e^{q_2 hs}ds < 1,\, t \geq t_0,\, u > 0\Big\},$$

$$\widetilde{\Theta}_{11} = M^{p+1}e^{-pht_0}\chi(p + q_1 + q_2),$$

$$\widetilde{\Theta}_{21} = M^{q_1+1}e^{-q_1 ht_0}\gamma_1(p + q_1 + q_2)\sum_{i=1}^m \|K_i(C_0 + \Delta C)\|,$$

$$\widetilde{\Theta}_{22} = M^{q_2+1}e^{-q_2 ht_0}\gamma_2(p + q_1 + q_2)\sum_{i=1}^m \|K_i(C_0 + \Delta C)\|,$$

$$\gamma_1 = \sum_{i=1}^m \gamma_{i1}, \quad \gamma_2 = \sum_{i=1}^m \gamma_{i2}.$$

Proof. The proof of Theorem 4.8 is carried out similarly to that of Theorem 4.7. Paper [90] presents the results of applying Theorem 4.7 to the mathematical model (4.93) for the interval values of the parameters and the initial conditions of motion. ◻

in which

$$A_0 = \begin{pmatrix} -\dfrac{R_{s0}}{L_d} & \dfrac{n_p L_q \omega_0}{L_d} & \dfrac{n_p L_q i_{q0}}{L_d} \\ -\dfrac{n_p L_d \omega_0}{L_q} & -\dfrac{R_{s0}}{L_q} & \left(-\dfrac{n_p L_d i_{d0}}{L_q} - n_p\Phi\right) \\ \dfrac{3n_p i_{d0}}{2J_0}(L_d - L_q) & \dfrac{3n_p i_{d0}}{2J_0}(L_d - L_q) + \dfrac{3n_p\Phi}{2} & 0 \end{pmatrix},$$

$$\Delta A = \begin{pmatrix} \dfrac{R_{s0} - R_s}{L_d} & 0 & 0 \\ 0 & \dfrac{R_{s0} - R_s}{L_q} & 0 \\ 0 & 0 & 0 \end{pmatrix}.$$

(3) the interval initial conditions are given as

$$x(t_0) = x_0 \in [\underline{x}_0, \overline{x}_0]. \tag{4.113}$$

Using Theorem 4.8, we establish sufficient conditions for the (λ, A, J)-stability of a solution $x = 0$ of the system (4.93) under the interval initial conditions (4.113). The stability conditions will be found in the form of constraints on the control matrix $K_0 = \begin{pmatrix} k_1 \\ k_2 \end{pmatrix}$, $K \in \mathbb{R}$, and the system parameters (4.93).

Since $G(0) = 0$, assumption D_{12} is satisfied. Consider the matrix $A_0 - BK_0C$ from assumption D_{13}. Denote by a_{ij} $(i, j = 1, 2, 3)$ the elements of the matrix A_0, while $a_{33} = 0$. Then, $A_0 - BK_0C$ is equal to

$$A_0 - BK_0C = \begin{pmatrix} a_{11} & a_{12} & a_{13} - k_1 \\ a_{21} & a_{22} & a_{23} - k_2 \\ a_{31} & a_{32} & 0 \end{pmatrix}.$$

Since the matrix $A_0 - BK_0C$ is constant, the fundamental matrix of the system from assumption D_{13} is equal to the exponential of $A_0 - BK_0C$; therefore, the condition (4.51) will be satisfied if and only if

$$-\alpha = \max_i Re\lambda_i(A_0 - BK_0C) < 0. \tag{4.114}$$

As is known, the fulfilment of the conditions (4.114) can be ensured by the fulfilment of the Routh–Hurwitz conditions for the characteristic polynomial of the matrix $A_0 - BK_0C$. For the characteristic polynomial of the third order

$$f(z) = z^3 + a_1 z^2 + a_2 z + a_3,$$

these conditions are

$$a_1 > 0, \quad a_1 a_2 - a_3 > 0, \quad a_3 > 0. \tag{4.115}$$

After calculating the characteristic polynomial for the matrix $A_0 - BK_0C$, the conditions (4.115) can be rewritten as

$$\begin{cases} -a_{11} - a_{22} > 0, \\ (-a_{11} - a_{22})(a_{11}a_{22} - a_{31}a_{13} + a_{31}k_1 - a_{32}a_{23} + a_{32}k_2 - a_{12}a_{21}) \\ + a_{12}a_{31}a_{23} - a_{12}a_{31}k_2 + a_{21}a_{32}a_{13} - a_{21}a_{32}k_1 - a_{31}a_{22}a_{13} \\ + a_{31}a_{22}k_1 - a_{32}a_{11}a_{23} + a_{32}a_{11}k_2 > 0, \\ -a_{12}a_{31}a_{23} + a_{12}a_{31}k_2 - a_{21}a_{32}a_{13} + a_{21}a_{32}k_1 + a_{31}a_{22}a_{13} \\ -a_{31}a_{22}k_1 + a_{32}a_{11}a_{23} - a_{32}a_{11}k_2 > 0. \end{cases} \tag{4.116}$$

The first inequality from (4.116) holds because $a_{11} = -\frac{R_{s0}}{L_d} < 0$, $a_{22} = -\frac{R_{s0}}{L_q} < 0$. Then, the system (4.116) can be rewritten as

$$PK_0 > Q, \tag{4.117}$$

where

$$P = \begin{pmatrix} -a_{11}a_{31} - a_{22}a_{31} & -a_{22}a_{32} - a_{12}a_{31} \\ -a_{21}a_{32} + a_{31}a_{22} & -a_{11}a_{31} + a_{32}a_{11} \end{pmatrix},$$

$$Q = \begin{pmatrix} a_{11}^2 a_{22} - a_{11}a_{31}a_{13} - a_{11}a_{12}a_{21} + a_{22}^2 a_{11} - a_{22}a_{32}a_{23} \\ -a_{22}a_{12}a_{21} - a_{12}a_{31}a_{23} - a_{21}a_{32}a_{13} \\ a_{12}a_{31}a_{23} + a_{21}a_{32}a_{13} - a_{31}a_{22}a_{13} - a_{32}a_{11}a_{23} \end{pmatrix}, \tag{4.118}$$

and the inequality is taken element by element.

Thus, if K_0 satisfies the inequality (4.117), then condition D_{14} is satisfied.

Consider assumption D_{14}. Let us estimate the function $G(x(t))$ in the norm

$$\|G(x(t))\| \leq \left| \frac{3n_p i_{q0}}{2J_0}(L_d - L_q)x_1 + \left(\frac{3n_p i_{d0}}{2J_0}(L_d - L_q) + \frac{3n_p \Phi}{2} \right) x_2 \right.$$

$$+ \left. \frac{3n_p}{2J_0}(L_d - L_q)x_1 x_2 \right| \leq \left| \frac{3n_p i_{q0}}{2J_0}(L_d - L_q)x_1 \right.$$

$$+ \left(\frac{3n_p i_{d0}}{2J_0}(L_d - L_q) + \frac{3n_p \Phi}{2} \right) x_2 \right| + \left| \frac{3n_p}{2J_0}(L_d - L_q)x_1 x_2 \right|$$

$$\leq \gamma_{11}\|x\| + \gamma_{12}\|x\|^2, \tag{4.119}$$

where

$$\gamma_{11} = \sqrt{\left(\frac{3n_p i_{q0}}{2J_0}(L_d - L_q) \right)^2 + \left(\frac{3n_p i_{d0}}{2J_0}(L_d - L_q) + \frac{3n_p \Phi}{2} \right)^2},$$

$$\gamma_{12} = \left| \frac{3n_p}{2J_0}(L_d - L_q) \right|.$$

Consider assumption D_{15}. Let us estimate $Y(x(t))$ in the norm

$$\|Y(x(t))\| \leq \|r\| \cdot \|x\|^2, \tag{4.120}$$

where $r = \left(\frac{n_p L_q}{L_d}, -\frac{n_p L_d}{L_q}, \frac{3n_p}{2J_0}(L_d - L_q) \right)^T$. From (4.119) and (4.120), it follows that the conditions D_{14} and D_{15} are satisfied for any values of the system parameters, while $i = 1$, $q_1 = 1$, $q_2 = 2$, and $p = 1$. Let us now find out the form of the inequalities (4.70) and (4.71) for system (4.93). Since $p = 1$, $q_1 = 1$, and $q_2 = 2$, the function $\tilde{\sigma}(u)$ becomes

$$\tilde{\sigma}(u) = \sup\left\{ t : \tilde{\Theta}u \int_{t_0}^t e^{hs} ds + \tilde{\Theta}_{22} u^2 \int_{t_0}^t e^{2hs} ds < 1, \, t \geq t_0 \right\},$$

where

$$\tilde{\Theta} = 4M^2 \chi e^{-ht_0} + 4\gamma_{11} M^2 |K| e^{-ht_0}, \quad \tilde{\Theta}_{22} = 4\gamma_{12} M^3 |K| e^{-2ht_0},$$

$$h = M|R_{s0} - R_s| \sqrt{\frac{1}{L_d^2} + \frac{1}{L_q^2}} - \alpha. \tag{4.121}$$

Since the function $\tilde{R}(u,t) = \tilde{\Theta}u \int_{t_0}^t e^{hs} ds + \tilde{\Theta}_{22} u^2 \int_{t_0}^t e^{2hs} ds$ on the left-hand side of the inequality in the definition of $\tilde{\sigma}(u)$ is increasing with respect to the variable $t \in [t_0, +\infty)$, we have that $\tilde{\sigma}(u)$ is equal to $t^*(u)$, where $t^*(u)$ is the solution of the equation $R(u,t) = 1$. Using simple calculations, we find that

$$t^*(u) = \lim_{s \to h} \ln \left(\frac{-\tilde{\Theta} + \sqrt{\tilde{\Theta}^2 + \tilde{\Theta}_{22}(2u\tilde{\Theta}e^{st_0} + \tilde{\Theta}_{22}u^2 e^{2st_0} + 2s)}}{\tilde{\Theta}_{22} u} \right)^{\frac{1}{s}}.$$

$$\tag{4.122}$$

Further, the function $\tilde{F}(u,t)$ from the condition of Theorem 4.8 becomes

$$\tilde{F}(u,t) = \frac{Mu e^{h(t-t_0)}}{\left(1 - \tilde{\Theta}u \int_{t_0}^t e^{hs} ds - \tilde{\Theta}_{22} u^2 \int_{t_0}^t e^{2hs} ds \right)^{\frac{1}{4}}}.$$

Now, the conditions for (λ, A, J)-stability of the motion of the system (4.93) under the interval initial conditions (4.116) are as follows:

(1) The inequality (4.118) is solvable with respect to the matrix K_0.
(2) The inequalities (4.117) are satisfied.
(3) Given λ, A, and T, the following inequalities hold:

  (a) $t^*(\lambda) > t_0 + T$;

(b)

$$\begin{cases} \dfrac{M\lambda e^{hT}}{\left(1 - \widetilde{\Theta}\lambda \displaystyle\int_{t_0}^{t_0+T} e^{hs}ds - \widetilde{\Theta}_{22}\lambda^2 \displaystyle\int_{t_0}^{t_0+T} e^{2hs}ds\right)^{\frac{1}{4}}} < A \\[20pt] h \geq 0, \end{cases}$$

$$\bigvee \begin{cases} \dfrac{M\lambda e^{hT}}{\left(1 - \widetilde{\Theta}\lambda \displaystyle\int_{t_0}^{t_0+T} e^{hs}ds - \widetilde{\Theta}_{22}\lambda^2 \displaystyle\int_{t_0}^{t_0+T} e^{2hs}ds\right)^{\frac{1}{4}}} < A \\[20pt] M\lambda < A, \\[6pt] h < 0, \end{cases}$$

where the constants $\widetilde{\Theta}$, $\widetilde{\Theta}_{22}$, and h are found using the formulas (4.121).

4.8   Comments and Bibliography

The method of integral inequalities developed in the qualitative theory of equations (see Bainov and Simeonov [12], Brauer [24], Lakshmikantham, Leela, and Martynyuk [66], and the bibliography therein) turned out to be applicable in this problem after some development (see Martynyuk [83]) based on a pseudo-linear representation of nonlinear integral inequalities (see Louartassi, Houssine, Mazoudi, and Alami [72] and N'Doye[124]).

Many works are devoted to the study of bilinear systems and their applications; see, for example, Echchatbi, Alami, and Bouaziz [37], España M. and Landau [38], Longchamp [71], Benallou, Mellichamp, and Seborg [16], and Mohler [57, 121, 122].

Sections 4.2–4.7 are based on the results of the papers by Babenko and Martynyuk [10, 11] and Martynyuk and Babenko [88, 89].

The problem of stability and stabilisation of the motion of affine systems attracts the attention of many specialists; see Sun and Guo [163], etc. In this case, the direct Lyapunov method and its modifications, determined by the formulation of the problem, are mainly used.

The development of control theory is a significant factor in the development of many areas of engineering and technology (see Cruz-Hernandez *et al.* [36], Lee and Markus [73], Mazko [109–112], Patan [134], Schumacher [157], Zecevic and Siljak [172]). The results obtained in this chapter have potential for application in the study of neural networks under interval initial conditions, as well as in the problem of meeting motions under interval initial conditions of motions.

Chapter 5

Stability and Boundedness of Solutions to Fractional-like Equations

5.1 Introduction

This chapter continues the development of integral methods for analysing solutions to equations of perturbed motion with a fractional-like derivative (F-LD) of the state vector. Here, we present the method of integral inequalities for estimating the norm of solutions of fractional-like equations and Lyapunov functions. On the basis of the estimates obtained, new sufficient conditions for various types of stability and boundedness of solutions of the systems of equations under consideration are established.

The chapter is organised according to the following plan.

Section 5.2 provides some information from mathematical analysis for functions with an F-LD.

Section 5.3 gives a physical interpretation of the F-LD of a continuous function.

In Section 5.4, for an inhomogeneous system with an F-LD of the state vector, estimates for the norm of solutions are established and some corollaries for specific systems are given.

In Section 5.5, we obtain estimates for the norm of solutions for nonlinear fractional-like systems under constant perturbations.

Section 5.6 contains an analysis of the Heyers–Ulam–Rassias stability of a fractional-like scalar equation.

Section 5.7 presents integral estimates of Lyapunov functions for quasi-linear fractional-like systems of equations.

Section 5.8 establishes sufficient conditions for various types of boundedness of motion of fractional-like equations and stability in the sense of Lagrange.

Section 5.9 is devoted to the problem of boundedness of solutions to the system of Hopfield equations with an F-LD of the state vector.

The final Section 5.10 provides comments and a brief bibliography of papers on this topic.

5.2 Preliminary Results

Let $\mathbb{R}_+ = [0, \infty)$, $\mathbb{R}^n$ be an n-dimensional Euclidean space, and $\Omega \subset \mathbb{R}^n$ be a bounded domain containing the origin.

For $q \in (0, 1]$ and a continuous function $x(t) : [t_0, \infty) \to \mathbb{R}$, we consider the generalised *F-LD* $D_t^q x(t)$ of order q (see [1, 60]).

Definition 5.1. Let a function $x : [t_0, \infty) \to \mathbb{R}$ be given. For any $q \in (0, 1]$, we define the expression $D_{t_0}^q(x(t))$ by the formula

$$D_{t_0}^q(x(t)) = \lim \left\{ \frac{[x(t + \theta(t - t_0))^{1-q}] - x(t)}{\theta}, \ \theta \to 0 \right\}.$$

The expression $D_{t_0}^q(x(t))$ is called the F-LD of order $0 < q \leq 1$ of the function $x(t)$. If $t_0 = 0$, then $D_{t_0}^q(x(t))$ becomes

$$D_{t_0}^q(x(t)) = \lim \left\{ \frac{[x(t + \theta t^{1-q})] - x(t)}{\theta}, \ \theta \to 0 \right\}.$$

For $t_0 = 0$, we write $D_0^q(x(t)) = D^q(x(t))$.

If $D^q(x(t))$ exists on $(0, b)$, then $D^q(x(0)) = \lim_{t \to 0^+} D^q(x(t))$.

If the F-LD of order q of a function $x(t)$ exists on (t_0, ∞), then we say that $x(t)$ is q-differentiable on (t_0, ∞).

Remark 5.1. In contrast to many definitions of the fractional derivative of a continuous (or absolutely differentiable) function, Definition 5.1 uses a limit rather than an integral, which significantly changed its properties.

The following assertion holds.

Lemma 5.1. *Let* $q \in (0, 1]$, *and let* $x(t)$ *and* $y(t)$ *be* q-differentiable *functions at* $t > 0$.

Then, the following relations are true:

(a) $D_{t_0}^q(ax(t) + by(t)) = aD_{t_0}^q(x(t)) + bD_{t_0}^q(y(t))$ *for all* $a, b \in \mathbb{R}$;
(b) $D_{t_0}^q(t^p) = p(t - t_0)^{1-q} t^{p-1}$ *for all* $p \in \mathbb{R}$;

(c) $D_{t_0}^q(x(t)y(t)) = x(t)D_{t_0}^q(y(t)) + y(t)D_{t_0}^q(x(t))$;

(d) $D_{t_0}^q\left(\dfrac{x(t)}{y(t)}\right) = \dfrac{y(t)D_{t_0}^q(x(t)) - x(t)D_{t_0}^q(y(t))}{y^2(t)}$;

(e) $D_{t_0}^q(x(t)) = 0$ *for any function* $x(t) = \lambda$.

Remark 5.2. For all known fractional derivatives (see [59, 139]), including the Riemann–Liouville fractional derivative

$$D_{t_0}^r x(t) = \frac{1}{\Gamma(n-q)}\frac{d^n}{dt^n}\int_{t_0}^t \frac{x(s)}{(t-s)^{q-n+1}}ds,$$

where $n - 1 < q < r$ and $\Gamma(z) = \int_0^\infty e^{-t}t^{2-1}dt$ is the gamma function, and the Euler and Caputo derivative [29]

$$D_{t_0}^c x(t) = \frac{1}{\Gamma(n-q)}\int_{t_0}^t \frac{x^n(s)}{(t-s)^{q-n+1}}ds,$$

statements (a)–(e) do not hold, except for statement (e) for the fractional Caputo derivative. The reason for this is the use of the integral in the definitions of the fractional derivative.

The fractional-like integral of order $0 < q \le 1$ is introduced by the formula

$$I_{t_0}^q x(t) = \int_{t_0}^t x(s)d_q s = \int_{t_0}^t (s-t_0)^{q-1}x(s)ds,$$

where the integral is understood in the Riemann sense.

The following assertion holds.

Lemma 5.2. *Let a function* $x(t) : (t_0, \infty) \to \mathbb{R}$ *be q-differentiable for* $0 < q \le 1$. *Then, for all* $t > t_0$, *the relation*

$$I_{t_0}^q(D_{t_0}^q x(t)) = x(t) - x(t_0)$$

is fulfilled.

Further, some of the above results are applied in the analysis of the equations of perturbed motion with an F-LD of the state vector of the system.

5.3 Interpretation of the Fractional-like Derivative

The use of the limit in Definition 5.1, instead of the integral used in the classical definitions of the fractional derivative of Riemann–Liouville,

Caputo, and others, allows us to give the following *physical interpretation* of the *F-LD*.

Let the point P move along a straight line to $\mathbb{R}_+$. For times $t_1 = t$ and $t_2 = t + \theta(t - t_0)^{1-q}$, where $\theta > 0$ and $0 < q \leq 1$, denote by $S(t_1)$ and $S(t_2)$ the path traversed by the point P during the times t_1 and t_2, respectively.

It is obvious that the ratio

$$\frac{S(t_2) - S(t_1)}{t_2 - t_1} = \frac{S(t + \theta(t - t_0)^{1-q}) - S(t)}{\theta(t - t_0)^{1-q}} = v_{avr}(t)$$

is the q-average velocity of the point P during the time $\theta(t - t_0)^{1-q}$.

Consider

$$D_{t_0}^q(S(t)) = \lim \left\{ \frac{S(t + \theta(t - t_0)^{1-q}) - S(t)}{\theta}, \ \theta \to 0 \right\}$$

$$= \lim \left\{ \frac{S(t + \theta(t - t_0)^{1-q}) - S(t)}{\theta(t - t_0)^{1-q}} (t - t_0)^{1-q}, \ \theta \to 0 \right\}$$

$$= \frac{dS}{dt} (t - t_0)^{1-q} = v_{inst}(t),$$

where $\frac{dS}{dt}$ is the usual instantaneous velocity of the point P.

For $q = 1$, this is the usual instantaneous speed of the point P at any time t on $\mathbb{R}_+$. For $0 < q < 1$, this is the q-instantaneous velocity of the point P for any value of t on $\mathbb{R}_+$.

Thus, the physical meaning of the F-LD is the q-instantaneous rate of change of the state vector of the mechanical or other natural system under consideration.

This is the answer to the question posed by Professor T. A. Burton, which he formulated when discussing the paper [95]: "What is the physical meaning of the F-LD of a continuous function?"

5.4   Integral Estimates for the Norms of Solutions

We consider a fractional-like system of equations of perturbed motion:

$$D_{t_0}^q x(t) = g(t, x(t)) + f(t), \tag{5.1}$$

$$x(t_0) = 0, \tag{5.2}$$

where $x \in \mathbb{R}^n$, $g \in C^q(\mathbb{R}_+ \times \mathbb{R}^n, \mathbb{R}^n)$, $f : \mathbb{R}_+ \to \mathbb{R}$, and $t > t_0$. It is assumed that the initial problem (5.1)–(5.2) has a solution $x(t, t_0, 0) \in C^q(J \times \mathbb{R}_+ \times \mathbb{R}^n, \mathbb{R}^n)$ for all $t \in J$, where $J \subset \mathbb{R}_+$ is an open interval.

Remark 5.3. An initial value problem similar to (5.1)–(5.2) with the fractional Caputo derivative was considered in the monograph [27, p. 128].

The following assertion holds.

Lemma 5.3. *If the above conditions are satisfied for a fractional-like system,* (5.1), *then it transforms into the equation*

$$x(t) = \int_{t_0}^{t} \frac{f(s)}{(s - t_0)^{1-q}} ds + \int_{t_0}^{t} \frac{g(s, x(s))}{(s - t_0)^{1-q}} ds \qquad (5.3)$$

for all $t \in J$.

Proof. Since $x(t)$, $g(t, x(t))$, and $f(t)$ are continuous functions, the integrals $I_{t_0}^q(g(t, x(t)))$ and $I_{t_0}^q(f(t))$ exist, and $I_{t_0}^q(D_{t_0}^q x(t))$ exists for all $t \in J$. In this case, Lemma 5.2 implies that

$$I_{t_0}^q(D_{t_0}^q x(t)) = x(t) - x(t_0). \qquad (5.4)$$

Applying the operator $I_{t_0}^q$ to the system of equations (5.1) and taking into account the initial conditions (5.2), we obtain

$$x(t) = I_{t_0}^q(g(s, x(s) + f(s)))$$

for all $t > t_0$. This proves Lemma 5.3. $\qquad\qquad\qquad\qquad\qquad\qquad \square$

Let us make some assumptions:

H_1. $F(t) = \int_{t_0}^{\infty} \frac{f(s)}{(s - t_0)^{1-q}} ds < +\infty$ uniformly in $t_0 \in \mathbb{R}_+$;

H_2. there exists a positive constant $k > 0$ such that $\|g(t, x)\| \le k\|x\|$ for all $(t, x) \in J \times B_r$, where $B_r = \{x \in \mathbb{R} : \|x\| \le r\}$.

Let us show that the following statement holds.

Theorem 5.1. *Let the conditions H_1 and H_2 be satisfied for a fractional-like system,* (5.1). *Then, for the solutions of the system* (5.1), *we have the estimate*

$$\|x(t)\| < B \exp\left(k \frac{(t - t_0)^q}{q}\right) \qquad (5.5)$$

for all $t > t_0$ and $0 < q \le 1$.

Proof. Under condition H_1 of Theorem 5.1, there exists a positive constant $B > 0$ such that

$$\|F(t)\| \leq B \tag{5.6}$$

for all $t \in J \subset \mathbb{R}_+$. Taking into account the conditions H_2 and the estimate (5.6), as well as the relation (5.4), we arrive at the integral inequality

$$\|x(t)\| \leq B + \int_{t_0}^{t} k(s - t_0)^{q-1}\|x(s)\|ds. \tag{5.7}$$

Denote $v(s) = k(s - t_0)^{q-1}$. Then,

$$\|x(t)\| \leq B + \int_{t_0}^{t} v(s)\|x(s)\|ds.$$

Applying Lemma 1.2 to this inequality, we obtain

$$\|x(t)\| \leq B \exp\left(\int_{t_0}^{t} v(s)ds\right) = B \exp\left(k\frac{(t - t_0)}{q}\right)^q$$

for all $t \in J$.

This proves Theorem 5.1. $\qquad\qquad\qquad\qquad\qquad\qquad\qquad\qquad\square$

Corollary 5.1. *If in a fractional-like system, (5.1), the condition H_1 is satisfied and the vector function $g(t, x) = A(t)x$, where $A(t)$ is an $n \times n$ matrix with continuous elements on any finite interval, then the solution $x(t)$ satisfies the estimate (5.5) with a constant $k = \max(\|A(t)\| : t \in J)$.*

Corollary 5.2. *If in a fractional system, (5.1), the condition H_1 is satisfied and the vector function $g(t, x) = Cx$, where C is an $n \times n$ constant matrix, then the relation*

$$x(t) = \int_{t_0}^{t} \exp\left(C\frac{(t - t_0)^q}{q}\right) \exp\left(-C\frac{(s - t_0)^q}{q}\right) f(s)(s - t_0)^{1-q}ds$$

is the solution of the system (5.1) under the initial conditions (5.2).

5.5   System under Permanent Disturbances

We consider a *fractional-like system* of equations under *permanent perturbations*

$$D_{t_0}^q x(t) = g(t, x(t)) + r(t, x(t)), \tag{5.8}$$

$$x(t_0) = 0, \tag{5.9}$$

where $x \in \mathbb{R}^n$, $g \in C^q(\mathbb{R}_+ \times \mathbb{R}^n, \mathbb{R}^n)$, $r \in C(\mathbb{R}_+ \times \mathbb{R}^n, \mathbb{R}^n)$, and $r(t,0) \neq 0$ for all $t \in \mathbb{R}_+$. Suppose that a solution $x(t)$ of the initial problem (5.8)–(5.9) exists on the interval J.

Lemma 5.4. *Let the system* (5.8) *satisfy the above conditions. Then, the initial problem* (5.8)–(5.9) *transforms into the equation*

$$x(t) = \int_{t_0}^{t} \frac{g(s, x(s))}{(s - t_0)^{1-q}} ds + \int_{t_0}^{t} \frac{r(s, x(s))}{(s - t_0)^{1-q}} ds \qquad (5.10)$$

for all $t \in J$.

Proof.   The proof of Lemma 5.4 is similar to that of Lemma 5.3.   □

Let us make the following assumptions about the components of the right-hand side of the system (5.8).

H_3. For constant perturbations $r(t, x(t))$, there exists a continuous function $h(t)$ such that $\|r(t, x)\| \le h(t)$ for all $(t, x) \in J \times B_r$.

H_4. $H(t) = \int_{t_0}^{\infty} \frac{h(s)}{(s - t_0)^{1-q}} ds < +\infty$ uniformly in $t_0 \in \mathbb{R}_+$ and $0 < q \le 1$.

H_5. There exists a continuous function $m(t)$ such that $\|g(t, x)\| \le m(t)\|x\|$ for all $(t, x) \in J \times B_r$.

Let us obtain an estimate of the solutions of the system (5.8) under zero initial conditions, i.e., under the conditions of the beginning of the motion from the state of equilibrium.

Theorem 5.2. *Assume that the system* (5.8) *satisfies the conditions H_3–H_5, then the norm of the solution $x(t)$ satisfies the estimate*

$$\|x(t)\| \le H(t) + \int_{t_0}^{t} m(r)H(r) \exp\left(\int_{r}^{t} m(s)(s - t_0)^{q-1} ds\right) dr \qquad (5.11)$$

for all $t \in J$ and $0 < q \le 1$.

Proof.   From the relation (5.10), under assumptions H_3–H_5, it is easy to obtain the integral inequality

$$\|x(t)\| \le H(t) + \int_{t_0}^{t} m(s)\|x(s)\|(s - t_0)^{q-1} ds \qquad (5.12)$$

for all $t \in J$.

Assuming in the inequality (5.12), $v(s) = m(s)(s - t_0)^{q-1}$, we arrive at the integral inequality

$$\|x(t)\| \leq H(t) + \int_{t_0}^{t} v(s)\|x(s)\|ds, \quad t > t_0. \tag{5.13}$$

Applying Theorem 5.2 from the monograph [66] to this inequality, we obtain the estimate

$$\|x(t)\| \leq H(t) + \int_{t_0}^{t} [v(s)H(s)] \exp\left(\int_{s}^{t} v(\xi)d\xi\right) ds \tag{5.14}$$

for all $t > t_0$. This corresponds to the estimate (5.11). This proves Theorem 5.2. □

Corollary 5.3. *If in a fractional-like system, (5.8), the vector function $g(t, x) = A(t)x$, where $A(t)$ is an $n \times n$ matrix with continuous elements on any finite interval, then the estimate (5.14) is*

$$\|x(t)\| \leq H(t) + k \exp\left(k \frac{(t - t_0)^q}{q}\right)$$

$$\times \int_{t_0}^{t} H(s) \exp\left(-k \frac{(s - t_0)^q}{q}\right)(s - t_0)^{q-1}ds$$

for all $t \in J$ and $0 < q \leq 1$. Here, $k = \max(\|A(t)\|) : t \in J$.

Corollary 5.4. *If the function $H(t)$ in the inequality (5.11) is nondecreasing for all $t > t_0$, then the estimate (5.14) becomes*

$$\|x(t)\| \leq H(t) \exp\left(\int_{t_0}^{t} v(s)ds\right) = H(t) \exp\left(\int_{t_0}^{t} m(s)(s - t_0)^{q-1}ds\right)$$

for all $t > t_0$ and $0 < q \leq 1$.

5.6   Heyers–Ulam–Rassias Stability

We consider a scalar equation with an F-LD of the state function

$$D_{t_0}^q x(t) = g(t, x) + r(t, x), \tag{5.15}$$

$$x(t_0) = 0, \tag{5.16}$$

where $g \in C([t_0, \infty) \times \mathbb{R}, \mathbb{R})$, $r \in C([t_0, \infty) \times \mathbb{R}, \mathbb{R})$, $r(t, 0) \neq 0$ for all $t \in [t_0, \infty)$. For the equation (5.1), we obtain the Heyers–Ulam–Rassias (H.U.R) *stability* conditions in the sense of the following definition.

Definition 5.2. (cf. [173]) A fractional-like equation, (5.15) is H.U.R-stable if for any $\varepsilon > 0$ and for a solution $x_\varepsilon(t) : J \to \mathbb{R}_+$ satisfying the inequality

$$|D_{t_0}^q x_\varepsilon(t) - g(t, x_\varepsilon(t)) - r(t, x_\varepsilon(t))| \leq \varepsilon, \tag{5.17}$$

there exists a solution $x(t)$ of the equation (5.1) and a constant $\Theta(\varepsilon) > 0$ such that

$$|x(t) - x_\varepsilon(t)| \leq Q(\varepsilon) \quad \text{for all} \quad t \in J. \tag{5.18}$$

Let us introduce the notation

$$H(t) = \frac{k - \varepsilon}{q}(t - t_0)^q, \quad \text{where } 0 < \varepsilon < k < +\infty;$$

$$v(s) = \frac{L}{(s - t_0)^{1-q}}, \quad 0 < q \leq 1, \ L > 0,$$

and show that the following assertion holds.

Theorem 5.3. *Assume that the following conditions are satisfied for the fractional-like differential equation* (5.15):

(1) *there is a constant $L > 0$ such that*

$$|g(t, x) - g(t, x_\varepsilon)| \leq L|x - x_\varepsilon|$$

for all $(t, x) \in D \subset \mathbb{R}_+ \times \mathbb{R}, (t, x_\varepsilon \in D)$;

(2) *there is a constant $k > 0$ such that*

$$|r(t, x) - r(t, x_\varepsilon)| \leq k$$

for all $(t, x) \in D, (t, x_\varepsilon) \in D$;

(3)

$$\sup_{t \geq t_0} \left(H(t) + \int_{t_0}^t [v(s)H(s)] \right) \exp\left(\int_s^t v(\xi)d\xi \right) ds < +\infty.$$

Then, the fractional-like differential equation (5.15) *is H.U.R-stable.*

Proof. It follows from the equation (5.15) that the fractional-like differential equation (5.15) becomes

$$x(t) = \int_{t_0}^t \frac{g(s, x(s))}{(s - t_0)^{1-q}}ds + \int_{t_0}^t \frac{r(s, x(s))}{(s - t_0)^{1-q}}ds, \quad t \geq t_0. \tag{5.19}$$

Next, consider the inequality (5.16), from which it follows that

$$D_{t_0}^q x_\varepsilon(t) - g(t, x_\varepsilon(t)) - r(t, x_\varepsilon(t)) \le \varepsilon \qquad (5.20)$$

or

$$D_{t_0}^q x_\varepsilon(t) - g(t, x_\varepsilon(t)) - r(t, x_\varepsilon(t)) \ge -\varepsilon. \qquad (5.21)$$

The inequality (5.20) implies that

$$x_\varepsilon(t) \le \int_{t_0}^t \frac{g(s, x_\varepsilon(s))}{(s - t_0)^{1-q}} ds + \int_{t_0}^t \frac{r(s, x_\varepsilon(s))}{(s - t_0)^{1-q}} ds + \int_{t_0}^t \frac{\varepsilon}{(s - t_0)^{1-q}} ds$$

$$= \int_{t_0}^t \frac{g(s, x_\varepsilon(s))}{(s - t_0)^{1-q}} ds + \int_{t_0}^t \frac{r(s, x_\varepsilon(s))}{(s - t_0)^{1-q}} ds + \frac{\varepsilon(t - t_0)^q}{q}. \qquad (5.22)$$

From the inequalities (5.19)–(5.21), we find the estimate

$$x(t) - x_\varepsilon(t) \le \int_{t_0}^t \frac{(g(s, x(s)) - g(s, x_\varepsilon(s)))}{(s - t_0)^{1-q}} ds$$

$$+ \int_{t_0}^t \frac{(r(s, x(s)) - r(s, x_\varepsilon(s)))}{(s - t_0)^{1-q}} ds - \frac{\varepsilon(t - t_0)^q}{q}. \qquad (5.23)$$

Taking into account conditions (1) and (2) of Theorem 5.3, from the inequality (5.22), we obtain

$$|x(t) - x_\varepsilon(t)| \le \int_{t_0}^t \frac{L|x(s) - x_\varepsilon(s)|ds}{(s - t_0)^{1-q}} + \int_{t_0}^t \frac{k ds}{(s - t_0)^{1-q}} - \frac{\varepsilon(t - t_0)^q}{q}. \qquad (5.24)$$

Taking into account the designations of the functions $H(t)$ and $v(s)$, from the inequality (5.23), we find

$$|x(t) - x_\varepsilon(t)| \le H(t) + \int_{t_0}^t v(s)|x(s) - x_\varepsilon(s)|ds. \qquad (5.25)$$

Applying to the inequality (5.24) Theorem 5.3 from the monograph [66], we get the estimate

$$|x(t) - x_\varepsilon(t)| \le H(t) + \int_{t_0}^t [v(s)H(s)] \exp\left(\int_s^t v(\xi)d\xi\right) ds \qquad (5.26)$$

for all $t \in J$.

Under condition (3) of Theorem 5.3, there exists $Q(\varepsilon) > 0$ such that $|x(t) - x_\varepsilon(t)| \le Q(\varepsilon)$ for all $t \in J$. This proves Theorem 5.3. □

Remark 5.4. The function $H(t)$ in the inequality (5.24) is positive and nondecreasing for all $t \in J$.

Therefore, the estimate (5.25) is transformed to the following:

$$|x(t) - x_\varepsilon(t)| \le H(t) \exp\left(\int_{t_0}^t \upsilon(s)ds\right), \ t \in J.$$

It follows from this estimate that condition (3) of Theorem 5.3 can be replaced by

$$\sup_{t \ge t_0} \left(H(t) \exp\left(\frac{L}{q}(t - t_0)^q\right)\right) < +\infty.$$

Moreover, the assertion of Theorem 5.3 holds true.

5.7 Integral Estimates for Lyapunov Functions

The following system of equations of perturbed motion with an F-LD of the state vector is considered:

$$D_{t_0}^q x(t) = A(t)x + g(t, x), \tag{5.27}$$

$$x(t_0) = x_0, \tag{5.28}$$

where $x \in \mathbb{R}^n$, $A(t)$ is an $n \times n$ matrix with continuous elements on any finite interval, $g \in C(\mathbb{R}_+ \times \mathbb{R}^n, \mathbb{R}^n)$, and $t > t_0$. It is assumed that the solution $x(t, t_0, x_0) \in C^q(\mathbb{R}_+ \times \mathbb{R}^n, \mathbb{R}^n)$ for all $t \in \mathbb{R}_+$ and $0 < q \le 1$.

Along with the estimates of the norm of solutions to a system of the form (5.27), it is of interest to obtain an estimate of the Lyapunov function on the solutions of a system of equations with an F-LD of the state vector and to obtain conditions for bounded motion.

Together with the system of equations (5.27), we consider a positive definite function $V(t, x) \in C^q(\mathbb{R}_+ \times \mathbb{R}^n, \mathbb{R}_+)$.

The *F-LD* of the *Lyapunov function* $V(t, x)$ is defined by the formula

$$D_t^q V(t, x)$$

$$= \limsup\left\{\frac{V(t + \theta(t - t_0)^{1-q}, \ x + \theta(t - t_0)^{1-q}(A(t)x + g(t, x))) - V(t, x)}{\theta}\right.$$

$$\left. : \theta \to 0^+\right\},$$

where $0 < q \le 1$.

The following assertion holds.

Lemma 5.5. *Let the system* (5.27) *have continuous functions* $a, b \in C(\mathbb{R}_+, \mathbb{R})$, *and an F-LD of the function* $V(t, x)$ *satisfies the estimate*

$$D_{t_0}^q V(t, x) \le a(t) V(t, x) + b(t). \qquad (5.29)$$

Then, on the solutions of the system (5.27), *the variation of the function* $V(t, x(t))$ *is estimated by the inequality*

$$V(t, x(t)) \le V(t_0, x_0) \exp\left[\int_{t_0}^t a(s) d_q s\right] + \int_{t_0}^t \exp\left[\int_\tau^t a(s) d_q s\right] b(\tau) d_q \tau \qquad (5.30)$$

for all $t \in [t_0, \infty)$.

In the inequality (5.30), and further in the text, $d_q s \equiv t^{q-1} ds$.

Proof. In view of the relation for the q-derivative of the product of two functions, we have

$$D_{t_0}^q \left\{ V(t, x(t)) \exp\left[-\int_{t_0}^t a(s) d_q s\right] \right\}$$

$$= \left\{ D_{t_0}^q V(t, x(t)) - a(t) V(t, x(t)) \right\} \exp\left[-\int_{t_0}^t a(s) d_q s\right] \qquad (5.31)$$

for all $t \ge t_0$. Integrating both sides of the equality (5.31), we obtain the estimate

$$V(t, x(t)) \exp\left[-\int_{t_0}^t a(s) d_q s\right] - V(t_0, x(t_0))$$

$$= \int_{t_0}^t \left[D_{t_0}^q V(\tau, x(\tau)) - a(\tau) V(\tau, x(\tau))\right] \exp\left[-\int_{t_0}^\tau a(s) d_q s\right] d_q \tau$$

$$\le \int_{t_0}^t b(\tau) \exp\left[\int_\tau^{t_0} a(s) d_q s\right] d_q \tau. \qquad (5.32)$$

This implies the estimate (5.30).

Let us show that the following statement holds for the system (5.27). □

Lemma 5.6. *Let the system* (5.27) *have functions* $a, b \in C^q(\mathbb{R}, \mathbb{R}_+)$ *and* $V(t, x(t))$ *be a function such that*

$$V(t, x(t)) \le b(t) + \int_{t_0}^t a(\tau) V(\tau, x(\tau)) d_q \tau$$

for all $t \in [0, \infty)$.

Then, the estimate

$$V(t, x(t)) \le b(t) + \int_{t_0}^{t} \exp\left[\int_{\tau}^{t} a(s)d_q s\right] a(\tau)b(\tau)d_q \tau \tag{5.33}$$

holds for all $t \in [0, \infty)$.

Proof. Denote

$$z(t) = \int_{t_0}^{t} a(\tau)V(\tau, x(\tau))d_q \tau.$$

Then, $z(t_0) = 0$, and the relation

$$D_{t_0}^q z(t) = a(t)V(t, x(t)) \le a(t)[b(t) + z(t)] = a(t)b(t) + a(t)z(t) \tag{5.34}$$

holds for all $t \in [0, \infty)$.

The inequality (5.34) implies that

$$z(t) \le \int_{t_0}^{t} \exp\left[\int_{\tau}^{t} a(s)d_q s\right] a(\tau)b(\tau)d_q \tau, \tag{5.35}$$

and hence we obtain the assertion of Lemma 5.6 since $V(t, x(t)) \le b(t) + z(t)$ for all $t \in [0, \infty)$. $\qquad\square$

Corollary 5.5. *Let the function $a(t) \ge 0$ in Lemma 5.6 be continuous for all $t \in [t_0, \infty)$ and $b(t) = k$, $k \in \mathbb{R}$ for all $t \in [t_0, \infty)$.*

If for all $t \in [t_0, \infty)$, the inequality

$$V(t, x(t)) \le k + \int_{t_0}^{t} a(s)V(s, x(s))d_q s \tag{5.36}$$

holds, then

$$V(t, x(t)) \le k \exp\left(\int_{t_0}^{t} a(s)d_q s\right) \tag{5.37}$$

for all $t \in [t_0, \infty)$.

Proof. From the estimate (5.33), under the condition $b(t) = k$, we obtain

$$V(t, x(t)) \le k + k \int_{t_0}^{t} \exp\left[\int_{\tau}^{t} a(s)d_q s\right] a(\tau)d_q \tau$$

$$= k\left(1 + \exp\left[\int_{t_0}^{\tau} a(s)d_q s\right] - \exp\left[\int_{\tau}^{t} a(s)d_q s\right]\right) = k \exp\left[\int_{t_0}^{t} a(s)d_q s\right] \tag{5.38}$$

for all $t \in [t_0, \infty)$. $\qquad\square$

5.8 Boundedness and Stability of Motion Conditions

Consider a fractional-like system, (5.27), and assume that the vector function

$$F(t,x) = A(t)x + g(t,x) \neq 0 \quad \text{for} \quad x = 0,$$

is defined and g is continuous on $\mathbb{R}_+ \times \mathbb{R}^n$ for $0 < q \leq 1$.

We introduce the following definition (cf. [67]).

Definition 5.3. The solution $x(t, t_0, x_0)$ of the fractional-like system (5.27) is:

- B_1 – *bounded* if there exists a constant $p > 0$ such that for any $t_0 \in \mathbb{R}_+$ and $x_0 \in \mathbb{R}^n$, the condition $\|x(t,t_0,x_0)\,| < p$ holds for all $t \geq t_0$, and $0 < q \leq 1$;
- B_2 – *equi-bounded* if for any $t_0 \in \mathbb{R}_+$ and $\delta > 0$, there exists $p = p(t_0, \delta)$ such that for the initial conditions $x_0 \in \mathbb{R}^n$, $\|x_0\| < \delta$, the estimate $\|x(t,t_0,x_0)\| < p$ follows for all $t \geq t_0$ and $0 < q \leq 1$;
- B_3 – *uniformly bounded* if p in B_2 does not depend on t_0 for all $0 < q \leq 1$;
- B_4 – *quasi-equi-extremely constrained* if there exists $\bar{p} > 0$ and for any $\delta_0 > 0$, there exists $\tau = \tau(t_0, \delta_0) > 0$ such that from $\|x_0\| < \delta_0$, $x_0 \in \mathbb{R}^n$, $\|x(t,t_0,x_0)\| < \bar{p}$ for all $t \geq t_0 + \tau$ and $0 < q \leq 1$;
- B_5 – *quasi-uniformly bounded with the boundary $\bar{p}$* if τ in B_4 does not depend on t_0 for all $0 < q \leq 1$;
- B_6 – *equi-limited with the boundary $\bar{p}$* if the conditions of definitions B_2 and B_4 are satisfied simultaneously for all $0 < q \leq 1$;
- B_7 – *eventually bounded* to the limit with the boundary $\bar{p}$ if the conditions of definitions of B_3 and B_5 are satisfied simultaneously for all $0 < q \leq 1$;
- B_8 – *equi-stable in the sense of Lagrange* if the conditions of defining B_2 are satisfied and, for any $t_0 \in \mathbb{R}_+$ and $\eta > 0$, there exist $\delta_0(t_0, \eta) > 0$ and $\tau = \tau(t_0, \eta) \in \mathbb{R}_+$ such that from the condition $\|x_0\| < \delta_0(t,\eta)$, it follows that $\|x(t,t_0,x_0)\| < \eta$ for all $t \geq t_0 + \tau$ and $0 < q \leq 1$;
- B_9 – *uniformly stable in the sense of Lagrange* if the conditions of definitions B_3 and B_8 are satisfied, in which the values δ_0 and τ do not depend on $t_0 \in \mathbb{R}_+$ for all $0 < q \leq 1$.

Next, the integral estimates of the norm of solutions of the system (5.27) are obtained on the basis of Lemmas 5.5 and 5.6. However, in the inequality (5.30), it is assumed that $a(t) < 0$ for all $t \in \mathbb{R}_+$.

Moreover, we assume that the function $V(t,x)$ is defined in $D_\rho^c = (t \in \mathbb{R}_+$ and $x \in S_\rho^c)$, where $S_\rho^c = (x \in \mathbb{R}^n : \|x\| \geq \rho)$, $\rho > 0$ and q-differentiable at all $0 < q \leq 1$.

The estimate (5.30) allows us to establish boundedness conditions for the solutions of the problem (5.27)–(5.28) in the form of the following statements.

Theorem 5.4. *Suppose there exists a q-differentiable Lyapunov function $V(t, x)$ defined in the domain D_ρ^c, which satisfies the following conditions:*

(1) *all conditions of Lemma 5.5 are satisfied;*
(2) *there are constants $0 < c_1 < c_2$ such that $c_1\|x\|^2 \leq V(t, x) \leq c_2\|x\|^2$ for all $(t, x) \in D_\rho^c$;*
(3) *for $x_0 \in \mathbb{R}^n$, for which $\|x_0\| < \alpha$, there is a value $p > 0$ such that*

$$\left(\exp\left[\int_{t_0}^t a(s)d_q s \right] + \frac{1}{c_1} \int_{t_0}^t \exp\left[\int_\tau^t a(s)d_q s \right] b(\tau)d_q\tau \right)^{1/2} < \frac{p}{k\alpha}$$

for all $t \geq t_0$ and $0 < q \leq 1$, where $k = \left(\frac{c_2}{c_1} \right)^{1/2}$.
Then, the solution $x(t, t_0, x_0)$ of the system (5.27) *is equi-bounded.*

Proof.   Let $x(t, t_0, x_0)$ be the solution of the initial problem (5.27)–(5.28) for $t_0 \in \mathbb{R}_+$ and $x_0 \in \mathbb{R}^n$, $\|x_0\| < \alpha$. Under condition (1) of Theorem 5.4, the function $V(t, x)$ has the estimate (5.30), which, under condition (2) of Theorem 5.4, becomes

$$\|x(t, t_0, x_0)\|$$

$$\leq \left(\frac{c_2}{c_1} \right)^{1/2} \|x_0\| \left(\exp\left[\int_{t_0}^t a(s)d_q s \right] + \frac{1}{c_1} \int_{t_0}^t \exp\left[\int_\tau^t a(s)d_q s \right] b(\tau)d_q\tau \right)^{1/2}$$
$$(5.39)$$

for all $t \geq t_0$.

Under condition (3) of Theorem 5.4, the inequality (5.39) implies that $\|x(t, t_0, x_0)\| < p$ for all $t \geq t_0$, i.e., the solution $x(t, t_0, x_0)$ of the problem (5.27)–(5.28) is equi-bounded. $\square$

Theorem 5.5. *Let for the fractional-like system* (5.27), *there exists a Lyapunov function $V(t, x)$ specified in Theorem 5.4 and:*

(1) *the conditions of Lemma* 5.5 *are satisfied;*
(2) *there exists a comparison function* $\psi_1(\|x\|)$ *such that* $\psi_1(\|x\|) \leq V(x,t)$, $\psi(r) \to \infty$ *as* $r \to \infty$;
(3) *for all* $t \geq t_0$, $0 < q \leq 1$, $x_0 \in \mathbb{R}^n$ *for which* $\|x_0\| < \alpha$, *there exists* $0 < m < \infty$ *such that the following inequality holds:*

$$\exp\left[\int_{t_0}^{t} a(s)d_q s\right] + \frac{1}{v_0}\int_{t_0}^{t}\exp\left[\int_{\tau}^{t} a(s)d_q s\right]b(\tau)d_q\tau \leq m < \infty, \quad (5.40)$$

where $v_0 = V(t_0, x_0)$.
Then, the solution $x(t, t_0, x_0)$ *of the system* (5.27) *is equi-bounded.*

Proof. Let $x(t, t_0, x_0)$ be the solution of the initial problem (5.27)–(5.28) with the initial conditions $t_0 \in \mathbb{R}_+$ and $x_0 \in S_\alpha$, where $S_\alpha = \{x \in \mathbb{R}^n : \|x\| < \alpha\}$, with $\alpha > \rho$. The continuity of the function $V(t,x)$ implies the existence of a constant $q(t_0, \alpha) > 0$ such that if $x_0 \in S_\alpha$, then $mV(t_0, x_0) \leq q(t_0, \alpha)$. Due to condition (2) of Theorem 5.5, we choose $p(t_0, \alpha) > 0$ so large that $\psi_1(p(t_0, \alpha)) > q(t_0, \alpha)$. Assume that there exists $t_1 > t_0$ such that $\|x(t_1, t_0, x_0)\| = p(t_0, \alpha)$. It follows from condition (3) of Theorem 5.5 that $V(t, x(t)) \leq mV(t_0, x_0)$ for all $t \geq t_0$. From here, it follows that $\psi(p(t_0, \alpha)) \leq q(t_0, \alpha)$, but this contradicts the choice of $t_1 > t_0$, for which $\|x(t_1)\| = p(t_0, \alpha)$. Hence, $\|x(t, t_0, x_0)\| < p(t_0, \alpha)$ for all $t \geq t_0$, i.e., the solution $x(t, t_0, x_0)$ is equi-bounded in the sense of the definition B_2. $\square$

Theorem 5.6. *Assume that for a fractional-like system,* (5.27), *there is a Lyapunov function* $V(t,x)$ *specified in Theorem 5.4, for which:*

(1) *there are comparison functions* $\psi_i, i = 1, 2$, *such that*

$$\psi_1(\|x\|) \leq V(t,x) \leq \psi_2(\|x\|),$$

where $\psi_1(r) \to \infty$ *as* $r \to \infty$ *and* $\psi_2 \in K$*-class;*
(2) *the conditions of Lemma* 5.6 *are satisfied and for all* $0 < q \leq 1$, *the inequality*

$$\exp\left[\int_{t_0}^{t} a(s)d_q s\right] + \frac{1}{v_0}\int_{t_0}^{t}\exp\left[\int_{s}^{t} a(\xi)d_q\xi\right]b(s)d_q s \leq 1$$

holds for all $t \geq t_0$ *uniformly in* $t_0 \in \mathbb{R}_+$.

Then, the solution $x(t, t_0, x_0)$ *of the system* (5.27) *is uniformly bounded in the sense of definition* B_3.

Proof. Under condition (1) of Theorem 5.6, choose $r(\alpha)$ independently of $t_0 \in \mathbb{R}_+$ so that $\psi_2(\alpha) < \psi_1(\rho)$. From the estimate (5.33)), under condition (2) of Theorem 5.6, we obtain the inequality $V(t, x(t)) \leq V(t_0, x_0)$ for all $t \geq t_0$. Further, applying the idea of the proof of Theorem 5.5, we obtain that $\|x(t, t_0, x_0)\| < \rho(\alpha)$ for all $t \geq t_0$ uniformly in $t_0 \in \mathbb{R}_+$. This proves Theorem 5.6. $\qquad\square$

Theorem 5.7. *Let the system* (5.27) *have a function* $V(t, x)$ *defined and* q*-differentiable in the domain* D_ρ^c *and, in addition,*

(1) *condition* (1) *of Theorem 5.6 is satisfied;*
(2) *the inequality* (5.29) *holds with the function* $a(t) < 0$ *for all* $t \geq t_0$.

Then, the solutions of the initial problem (5.27)–(5.28) *are uniformly limiting bounded in the sense of definition* B_7.

Proof. It follows from condition (1) of Theorem 5.7 that for $\|x\| \geq \rho_1$, $\rho_1 \geq \rho$, the function $V(t, x)$ is positive definite and radially unbounded. Next, we consider the function $V(t, x)$ in the domain D_ρ^c. Let the solutions $x(t) = x(t, t_0, x_0)$ have the initial conditions $t_0 \in \mathbb{R}_+$ and $\|x\| \leq \rho$. Let us show that under the conditions of Theorem 5.7, these solutions are quasi-uniformly limiting bounded.

For this, the interval $[t_0, T]$ of the existence of solutions $x(t)$ is divided into two sets: $[t_0, T] = T_1 + T_2$, where T_i $(i = 1, 2)$ is a set of the moments of time for which $\|x\| \leq \rho$ if $i = 1$ and $\|x\| > \rho$ if $i = 2$. If T_2 is an empty set, then $\|x\| \leq \rho$ and the statement of Theorem 5.7 is proved because $\|x\| \leq \rho$ for all $t \in [t_0, T]$. Assume that T_2 is not an empty set. From the fact that $\|x\| > \rho$ is open, it follows that the set T_2 is also open. Therefore, the set T_2 can be represented as a finite or countable sum of intervals, i.e., $T_2 = \bigcup_\sigma (t_\sigma, t_\beta)$, where $\beta = \beta(\sigma)$. It follows that if $t \in T_2$, then $t \in (t_\sigma, t_\beta)$ for some σ. From conditions (1) and (2) of Theorem 5.7, it follows that

$$\psi_1(\|x(t)\|) \leq V(t, x(t)) \leq V(t_\sigma, x(t_\sigma)) \leq \psi_2(\|x(t_\sigma)\|) = \psi_2(\rho) \qquad (5.41)$$

for $\|x(t_\sigma)\| = \rho$. The fact that $\psi_1 \in KR$-class implies that there exists $\overline{\rho} \geq \rho > 0$ such that that $\psi_2(r) > \psi_1(\rho)$ as $r \geq \overline{\rho}$. Hence, from the inequalities (5.41), we find that

$$\|x(t)\| < \overline{\rho} \quad \text{at} \quad t \in T_2. \qquad (5.42)$$

The inequality (5.42), together with the condition $\|x(t_0)\| \leq \rho$, leads to the conclusion that

$$\|x(t)\| < \overline{\rho} \quad \text{at} \quad t_0 \leq t < t_0 + T. \qquad (5.43)$$

This means that the solution $x(t)$ is infinitely extendable to the right, i.e., $T = \infty$, for all $t_0 \in \mathbb{R}_+$ and $\|x_0\| \leq \rho$, the estimate (5.43) holds, where $\overline{\rho}$ depends only on ρ.

Next, consider the solution $x(t, t_0, x_0)$ with the initial conditions $t_0 \in \mathbb{R}_+$ and $\|x(t_0)\| > \rho$. Let us show that in this case, there exists $t_1 > t_0$ such that the inequality (5.43) holds true at $t \geq t_1$. Note that if for $t_1 > t_0$, $\|x(t_1, t_0, x_0)\| \leq \rho$, then from the uniqueness of the solution, it follows that

$$x(t, t_0, x_0) = x(t, t_1, x(t_1, t_0, x_0));$$

therefore, based on the inequality (5.43), we have the estimate

$$\|x(t, t_0, x_0)\| < \overline{\rho} \quad \text{at} \quad t \geq t_1. \tag{5.44}$$

Indeed, let $\|x(t)\| > \rho$ for $t \geq t_0$. According to conditions (1) and (2) of Theorem 5.7, we have the estimates

$$\psi_1(\|x(t)\|) \leq V(t, x(t)) \leq V(t_0, x_0) \leq \psi_2(\|x_0\|) \tag{5.45}$$

for $t \geq t_0$ and $\psi_1(r) > \psi_2(\|x_0\|)$ for $r \geq \rho_- \geq \overline{\rho}$, where ρ_1 is large enough. The inequality (5.45) implies that $\|x(t)\| < \rho_1$ for all $t \geq t_0$. Here, ρ_1 depends only on x_0, i.e., any solution $x(t, t_0, x_0)$ with the initial conditions $t_0 \in \mathbb{R}_+$ and $\|x(t_0)\| > \rho$ is bounded on $[t_0, \infty)$ uniformly in $t_0 \in \mathbb{R}_+$ and extendable to the right indefinitely.

Let us show that there exists a moment $t_1 > t_0$ at which

$$\|x(t_1, t_0, x_0)\| = \rho. \tag{5.46}$$

We assume that

$$\rho < \|x(t, t_0, x_0)\| < \rho_1 \tag{5.47}$$

for $t_1 \geq t_0$ and compute

$$\inf_{\rho \leq \|x\| \leq \rho_1} (a(t)V(t, x) + b(t)) = \gamma > 0.$$

From condition (2) of Theorem 5.7, we find that

$$D_{t_0}^q V(t, x) \leq -\gamma \tag{5.48}$$

for all $t \geq t_0$ on the solutions $x(t, t_0, x_0)$ satisfying the estimate (5.47).

Applying the fractional-like integration operator to both sides of the inequality (5.48), we obtain the estimate

$$V(t, x(t)) \leq V(t_0, x_0) - \gamma \frac{(t - t_0)^q}{q} \tag{5.49}$$

for all $t \geq t_0$ and $0 < q \leq 1$.

It follows from the estimate (5.49) that for sufficiently large t for any $0 < q \leq 1$, the function $V(t, x(t))$ becomes negative in the range $(t, x(t)) \in D_\rho^c$. It contradicts the assumption about the properties of the function $V(t, x)$ specified in condition (1) of Theorem 5.7.

Therefore, the solution $x(t)$ for all $t \geq t_0$ cannot be within the boundaries (5.47), and for some $t_1 \geq t_0$, the relation (5.46) will be satisfied. From this, it follows that the estimate (5.43) is satisfied, and the solutions of the system (5.27) is quasi-uniformly bounded.

To estimate the value of t_1, we will consider the solution $x(t)$ in the domain $(t, x(t)) \in D_\rho^c$. From the inequality (5.49), we get

$$t_1 \leq t_0 + \left(q \frac{V(t_0, x_0) - V(t, x(t))}{\gamma} \right)^{\frac{1}{q}} \tag{5.50}$$

for all $0 < q \leq 1$.

It follows from conditions (1) of Theorem 5.7 that

$$\begin{aligned} V(t_0, x_0) \leq \psi_2(\|x_0\|) < \psi_2(\rho_1); \\ V(t, x(t)) \geq \psi_1(\|x(t)\|) > \psi_1(\rho). \end{aligned} \tag{5.51}$$

Given the estimates (5.51), the formula (5.50) becomes

$$t_1 \leq t_0 + T(x_0),$$

where

$$T(x_0) < \left(q \frac{\psi_2(\rho_1) - \psi_1(\rho)}{\gamma} \right)^{\frac{1}{q}}$$

for all $0 < q \leq 1$, $\rho_1 = \rho(\|x_0\|)$.

Since $T(x_0)$ does not depend on t_0, it follows that the solutions $x(t)$ of the system (5.27) are uniformly limiting bounded. Thus, Theorem 5.7 is proven. $\qquad \square$

Theorem 5.8. *If the conditions of Theorem 5.7 are satisfied with the value $\tau = \tau(\delta_0)$ for all $0 < q \leq 1$, then the solution $x(t, t_0, x_0)$ of the system (5.27) is quasi-uniformly bounded with the boundary $\bar{p} > 0$.*

Theorem 5.9. *If the conditions of Theorems 5.5 and 5.7 are satisfied simultaneously for all $0 < q \leq 1$, then the solution $x(t, t_0, x_0)$ of the system (5.27) is equi-limit bounded with the boundary $\bar{p} > 0$.*

Theorem 5.10. *If all the conditions of Theorems 5.6 and 5.8 are satisfied simultaneously for all $0 < q \leq 1$, then the solution $x(t, t_0, x_0)$ of the system (5.27) is uniformly limiting bounded with the boundary $\bar{p} > 0$.*

Further, the fractional-like system (5.27) is considered in the region of the values $(t, x) \in \mathbb{R}_+ \times D_H$, where $D_H = \{x \in R^n : \|x\| < H\}$, $H > 0$. We assume that $F(t, x) = 0$ for $x = 0$ and for all $t \in \mathbb{R}_+$. The following assertion holds.

Theorem 5.11. *Let there exist a Lyapunov function $V(t, x)$ defined and q-differentiable for all $(t, x) \in \mathbb{R}_+ \times D_H$, $0 < q \leq 1$ and, in addition:*

(1) *condition (2) of Theorem 5.5 is satisfied;*
(2) *the conditions of Lemma 5.5 are satisfied, and the inequality*

$$\exp\left[\int_{t_0}^t a(s)d_qs\right] + \frac{1}{v_0}\int_{t_0}^t \exp\left[\int_s^t a(\sigma)d_q\sigma\right] b(s)d_qs \leq 1$$

holds for all $t \geq t_0$;
(3) *there exists a comparison function $\psi_3(r)$ continuous on $[0, H]$ such that*

$$a(t)V(t, x) + b(t) \leq -\psi_3(\|x\|).$$

Then, the solution $x(t, t_0, x_0)$ of the system (5.27) is equi-stable in the sense of Lagrange.

Proof. Under conditions (1) and (2) of the theorem, the solution $x(t, t_0, x_0)$ is equi-limited. Under condition (1) of this theorem, for any $\xi > 0$, we have the estimate

$$\psi_1(\xi) \leq V(t, x) \quad \text{at} \quad t \in \mathbb{R}_+ \quad \text{and} \quad x \in \mathbb{R}^n$$

such that $\|x\| = \xi$. For a fixed $t_0 \in \mathbb{R}_+$, choose $\delta_0(t_0) > 0$ so that if $\|x_0\| < \delta_0(t_0, \xi)$, then $V(t_0, x_0) < \psi_1(\xi)$. This is possible due to the continuity of the function $V(t, x)$ and the fact that $V(t_0, 0) = 0$. Let there exist $t_1 > t_0$ such that for $\|x_0\| < \delta_0(t_0, \xi)$, the solution $x(t_1, t_0, x_0)$ reaches the boundary of the sphere $\|x\| = \xi$.

Under condition (2) of Theorem 5.11, the estimate $V(t, x(t)) \leq V(t_0, x_0)$ holds, and hence $\psi_1(\xi) \leq V(t_1, x(t_1)) \leq V(t_0, x_0) < \psi_1(\xi)$. The resulting contradiction shows that the assumption that $t_1 > t_0$ exists for $\|x(t_1, t_0, x_0)\| = \xi$ is not correct. Hence, the solution $x(t, t_0, x_0)$ remains in the region $\|x\| < \xi$ for all $t \geq t_0$.

Next, we show that under condition (3) of Theorem 5.11, for some value $t_1 > t_0$, the following relation is valid:

$$\|x(t_1, t_0, x_0)\| = \rho, \quad \rho < \xi. \tag{5.52}$$

Let, for all $t \geq t_0$, the solution $x(t, t_0, x_0)$ be in the range

$$\rho < \|x(t, t_0, x_0)\| < \xi. \tag{5.53}$$

Calculate the value

$$\gamma^* = \inf_{\rho \leq \|x\| \leq \xi} \psi_3(\|x\|),$$

and consider an estimate for the F-LD of the function $V(t, x)$:

$$D_{t_0}^q V(t, x) \leq -\gamma^* \quad \text{at} \ \ t \geq t_0. \tag{5.54}$$

From the inequality (5.54), it follows that

$$\psi_1(\|x\|) \leq V(t, x(t)) \leq V(t_0, x_0) - \gamma^* \frac{(t - t_0)^q}{q} \tag{5.55}$$

for all $t \geq t_0$ and $0 < q \leq 1$. From the estimate (5.55), it follows that for a sufficiently large t, the function $V(t, x(t))$ becomes negative for $(t, x(t)) \in S_\rho^c$, which contradicts condition (1) of Theorem 5.11. Therefore, there exists $t_1 > t_0$ for which the relation (5.52) will be satisfied. The estimate for the moment t_1 follows from the inequalities (5.55). Namely,

$$t_1(t_0, \xi) \leq t_0 + \left(q \frac{V(t_0, x_0) - \psi_1(\xi)}{\gamma^*} \right)^{\frac{1}{q}} \tag{5.56}$$

for all $0 < q \leq 1$ and the estimates (5.53). (5.55) implies that the solution $x(t, t_0, x_0)$ satisfies the estimate $\|x(t, t_0, x_0)\| < \xi$ for all $t \geq t_0 + t_1(t_0, \xi)$ and $0 < q \leq 1$ for $\|x_0\| < \delta_0(t_0, \xi)$. Thus, under the conditions of Theorem 5.11, the solution $x(t, t_0, x_0)$ is equi-bounded and quasi-equi-asymptotically stable, i.e., equi-stable in the sense of Lagrange. $\square$

Theorem 5.12. *Suppose that for a fractional-like system, (5.27), there exists the Lyapunov function $V(t, x)$ satisfying the conditions $(2)'$ and $(3)'$ of Theorem 5.12 and condition (3) of Theorem 5.11.*

Then, the solution $x(t, t_0, x_0)$ of system (5.27) is uniformly in $t_0 \in \mathbb{R}_+$ stable in the sense of Lagrange.

Proof. Under conditions $(2)'$ and $(3)'$ of Theorem 5.3, the solution $x(t, t_0, x_0)$ of the system (5.27) is uniformly bounded for all $0 < q \leq 1$. Therefore, there exists $\delta_0 > 0$ such that for $t_0 \in \mathbb{R}_+$ and $\|x_0\| < \delta_0$, the solution $x(t, t_0, x_0)$ satisfies $\|x(t, t_0, x_0)\| < H$ for all $t \geq t_0$. Moreover, for any $\varepsilon > 0$, there exists $\delta(\varepsilon) > 0$ such that if $t_0 \in \mathbb{R}_+$ and $\|x_0\| < \delta(\varepsilon)$, then $\|x(t, t_0, x_0)\| < \varepsilon$ for all $t \geq t_0$. Let us show that the solution $x(t, t_0, x_0)$ of the system (5.27) under the initial conditions $t_0 \in \mathbb{R}_+$ and $\|x_0\| < \delta_0$

will satisfy the estimate $\|x(t_1, t_0, x_0)\| < \delta(\varepsilon)$ for some $t_1 \in \mathbb{R}_+$. Let the estimate $\delta(\varepsilon) \leq \|x(t, t_0, x_0)\| < H$ be true for all $t \geq t_0$. From condition (3)$'$ of Theorem 5.12, it follows that

$$V(t, x(t)) \leq V(t_0, x_0) - \tilde{\gamma}\frac{(t - t_0)}{q} \qquad (5.57)$$

for all $t \geq t_0$, where $\tilde{\gamma} = \inf_{\delta(\varepsilon) \leq \|x\| \leq H} \psi_3(\|x\|)$. Since $V(t_0, x_0) \leq \psi_2(\|x_0\|) < \psi_2(\delta_0)$ and $V(t, x(t)) \geq \psi_1(\|x(t)\|) \geq \psi_1(\delta)$ then it is not difficult to obtain the estimate. Therefore,

$$t_1 \leq t_0 + T(x_0),$$

where

$$T(x_0) < \left(q\frac{\psi_2(\delta_0) - \psi_1(\delta)}{\tilde{\gamma}} \right)^{\frac{1}{q}}.$$

The value of $T(x_0)$ does not depend on $t_0 \in \mathbb{R}_+$; therefore, for $\|x_0\| < \delta_0$, the estimate $\|x(t, t_0, x_0)\| < \delta(\varepsilon)$ for all $t \geq t_0 + T(x_0)$. This proves that the solution $x(t)$ of the system (5.27) is uniformly stable in the sense of Lagrange. $\qquad \square$

5.9   Boundedness of Solutions of Hopfield Equations

Consider the system of *Hopfield equations* with an *F-LD* of the state vector

$$D_{t_0}^q x_i(t) = -b_i x_i + \sum_{j=1}^{N} A_{ij} G_j(x_j(t)) + V_i(t), \quad i = 1, 2, \ldots, N. \qquad (5.58)$$

Here, we use the designations

$$A_{ij} = T_{ij}/C_i, \; V_i(t) = I_i(t)/C_i \; \text{u} \; b_i = 1/\tau_i C_i,$$

where $C_i > 0$, $T_{ij} = 1/R_{ij}$, $R_{ij} \in \mathbb{R}$, $1/\tau_i = 1/R_i + \sum_{j=1}^{N} |T_{ij}|$, and $R_i > 0$, $I_i : \mathbb{R}_+ \to [0, \infty)$, for all $i = 1, 2, \ldots, n$, and continuous, $D_{t_0}^q x_i(t)$ is an F-LD of the neural network state vector, $G_i : \mathbb{R} \to (-1, 1)$, and the functions G_i are continuously differentiable and strictly monotonically increasing, i.e., $G_i(x_i') > G_i(x_i'')$ if and only if $x_i' > x_i''$, $x_i G_i(x_i) > 0$ for all $x_i \neq 0$ and $G_i(x_i) = 0$ for $x_i = 0$, $i = 1, 2, \ldots, N$.

In the system of equations (5.58), C_i denotes capacitance, R_{ij} is resistance, $G_i(x_i)$ is the nonlinear amplifier function, and $I_i(t)$ is an external input to the network.

We make the following assumptions about the fractional-like system (5.58):

A_1. External inputs are non-zero, i.e.,

$$U_i(t) \neq 0 \quad \text{for all} \ \ t \in \mathbb{R}_+, \quad i = 1, 2, \ldots, N.$$

A_2. There are constants $a_{ij} \in \mathbb{R}$, $i, j = 1, 2, \ldots, N$, such that

$$x_i A_{ij} G_j(x_j) \leq x_i a_{ij} x_j$$

for all $(x_i, x_j) \in \mathbb{R}^n$, $i, j = 1, 2, \ldots, N$.

A_3. There is a continuous function $a_1(t) : \mathbb{R} \to \mathbb{R}$ such that

$$\tilde{a}(t) = -\sum_{i=1}^{N} b_i + a_1(t) \neq 0$$

for all $t \in \mathbb{R}_+$.

If the fractional-like system (5.58) satisfies the assumptions A_1–A_3, then for the F-LD of the function

$$V(t, x) = \sum_{i=1}^{N} \frac{1}{2} \alpha_i(t) x_i^2,$$

there is an estimate,

$$D_{t_0}^q V(t, x) \leq 2\tilde{a}(t) V(t, x) + b(t), \tag{5.59}$$

where

$$b(t) = \sum_{i=1}^{N} \alpha_i(t) r_i U_i(t), \quad \alpha_i(t) \in C(\mathbb{R}_+, \mathbb{R}_+) \ \ \text{and} \ \ r_i \geq |x_i|$$

are some constants.

Let the function $V(t, x)$ be given. Then,

$$D_{t_0}^q V(t, x) = \sum_{i=1}^{N} \alpha_i(t) x_i D_{t_0}^q x_i(t)$$

$$= \sum_{i=1}^{N} \alpha_i(t) x_i \left[-b_i x_i + \sum_{j=1}^{N} A_{ij} G_j(x_j) + U_i(t) \right]. \tag{5.60}$$

Taking into account the assumptions A_1–A_3, from the relation (5.60), we obtain the estimate

$$D_{t_0}^q V(t, x) \leq 2V(t, x) \left(-\sum_{i=1}^{N} b_i + a_1(t) \right) + \sum_{i=1}^{N} \alpha_i(t) x_i U_i(t), \tag{5.61}$$

which implies the inequality (5.59).

Applying Lemma 5.5 to the inequality (5.59), we obtain an estimate for the function $V(t, x)$ on the solutions of the fractional-like system (5.58) in the form of (5.30), where the function $a(t) = 2\tilde{a}(t)$.

The sufficient conditions for various types of boundedness of solutions of the fractional-like model (5.58) of the Hopfield neutron network are obtained by the direct application of Theorems 5.5–5.12 to the inequality (5.59), with the function $a(t) = 2\tilde{a}(t)$.

5.10   Comments and Bibliography

Let us recall that the concept of the fractional derivative of a continuous function appeared in mathematical analysis following L'Hopital's question to Leibniz in 1695: how to understand the expression $\dfrac{d^n x}{dt^n}$ for $n = 1/2$?

The increased interest in fractional derivative equations in the past two decades (see Caputo [29], Kilbas *et al.* [59], Lakshmikantham *et al.* [65], Podlybny [139], and the bibliography therein) has motivated many researchers to create high-quality methods for dynamic analysis of equations of perturbed motion with the fractional derivative of the system state vector. The reason for this interest was the possibility of a more accurate description of processes in some models of real-world phenomena observed in theoretical mechanics, electronics, and continuum mechanics.

Section 5.2 adapts some results from the papers of Abdeljawad [1], and Khalil *et al.* [60].

Sections 5.3–5.6 are based on the results of the paper by Martynyuk and Martynyuk-Chernienko [93].

Section 5.7 adapts the results of the papers by Martynyuk [87] and Martynyuk and Stamova [95].

Section 5.8 is based on some results of the paper by Martynyuk *et al.* [97].

Some aspects of the qualitative theory of equations with an F-LD of the system state vector have been developed in the past few years (see Martynyuk *et al.* [96, 98], Martynyuk *et al.* [99], Zheng *et al.* [173], and the bibliography therein).

Chapter 6

Elements of Fractional-like Dynamics on the Time Scale

6.1 Introduction

In this chapter, an integral method for analysing dynamic equations with an F-LD state vector on the time scale is discussed.

New estimates of the variation of Lyapunov functions along solutions of fractional-like equations are based on integral inequalities on the time scale.

The obtained estimates are used to analyse various types of stability and boundedness of dynamic equations with a F-LD state vector.

The results of this chapter are presented according to the following plan.

Section 6.2 presents some general results from mathematical analysis on the time scale for functions with a fractional-like derivative.

In Section 6.3, a system of fractional-like equations of perturbed motion on the time scale is considered. The generalised derivative of the Lyapunov function on the solutions of the system is introduced, and research problems are formulated.

In Section 6.4, new estimates of the Lyapunov function on solutions of fractional-like dynamic equations are given.

Section 6.5 contains some applications of general estimates of Lyapunov functions to qualitative motion analysis. Namely, it considers the problems of:

— motion boundedness;
— motion of a system with a given settling time;
— practical stability of motion;
— motion stability in the sense of Lyapunov.

6.2   Elements of Fractional-like Analysis on a Time Scale

6.2.1   Δ-derivative on the time scale [20, 50]

Let us recall the accepted notations and definitions.

Let $\mathbb{R}_+ = [0, \infty)$, $\mathbb{R}^n$ be an n-dimensional Euclidean space and $\Omega \subset \mathbb{R}^n$ be a bounded area containing the origin. Besides:

$\mathbb{T}$ is a *time scale*, an arbitrary nonempty closed subset of real numbers;

$\sigma(t) = \inf\{s \in \mathbb{T} : s > t\}$ is a *forward-jump operator*;

$\rho(t) = \sup\{s \in \mathbb{T}, s < t\}$ is a *backward-jump operator*;

$$\mathbb{T}^\varkappa = \begin{cases} \mathbb{T}\backslash(\rho(\sup\mathbb{T}, \sup\mathbb{T}] & \text{if } \sup\mathbb{T} < \infty, \\ \mathbb{T} & \text{if } \sup\mathbb{T} = \infty; \end{cases}$$

$f^\sigma : \mathbb{T} \to \mathbb{R}, f^\sigma(t) = f(\sigma(t))\ \forall t \in \mathbb{T}$, i.e., $f^\sigma = f \circ \sigma$;

the *point* $t \in \mathbb{T}$ is *scattered to the right* if $\sigma(t) > t$;

the *point* $t \in \mathbb{T}$ is *scattered to the left* if $\rho(t) < t$;

the *point* $t \in \mathbb{T}$ is *dense on the right* if $t < \sup\mathbb{T}$ and $\sigma(t) = t$;

the *point* $t \in \mathbb{T}$ is *dense on the left* if $t > \inf\mathbb{T}$ and $\rho(t) = t$;

$\mu(t) = \sigma(t) - t$ is the *granularity function* of the scale $\mathbb{T}$;

C_r is a set of *regular functions*;

the function $f : \mathbb{T} \to \mathbb{R}$ is called regulated provided its right-sided limits exist (finite) at all right-dense points in $\mathbb{T}$ and its left-sided limits exist (finite) at all left-dense points in $\mathbb{T}$;

C_{rd} is the set of functions $f : \mathbb{T} \to \mathbb{R}$ continuous at the right-dense points of $\mathbb{T}$, for which the left-hand limit exists and is finite at the left-dense points of $\mathbb{T}$.

On the time scale $\mathbb{T}$, consider the function $f : [t_0, \infty) \to \mathbb{R}$, and suppose that $t \in \mathbb{T}^\varkappa$.

Definition 6.1. The function $f(t)$ admits a Δ-*derivative* with respect to t *on the time scale* if there exists a number $\Delta(f)(t)$ at which for any $\epsilon > 0$, there exists $\delta > 0$ such that, for all $s \in (t - \delta, t + \delta) \cap \mathbb{T}$, the following condition is fulfilled:

$$|[f(\sigma(t)) - f(s)] - \Delta(f)(t)[\sigma(t) - s]| \le \varepsilon|\sigma(t) - s|.$$

The expression f^Δ denotes the Δ-derivative of the function f on the time scale $\mathbb{T}$.

6.2.2   Fractional-like derivative on the time scale

Based on the concept of Δ-derivative on the time scale and fractional-like derivative, the concept of Δ-*fractional-like derivative* on *the time scale* is formed (see [17, 141]).

The *function* $f : \mathbb{T} \to \mathbb{R}$ is *q-regressive* on $\mathbb{T}$ if for any $q \in (0,1]$, the condition

$$1 + \mu(t)f(t)t^{q-1} \neq 0$$

is satisfied for all $t \in \mathbb{T}^\varkappa$.

Let $\mathcal{R}^q$ be the set of all q-regressive functions f on $\mathbb{T}$.

For some $t \in \mathbb{T}^\varkappa$ and $\delta > 0$, the δ-neighbourhood of the point t is defined as $\mathcal{N}_t : (t - \delta, t + \delta) \cap \mathbb{T}$.

Definition 6.2. Let the function $f : \mathbb{T} \to \mathbb{R}$, $t \in \mathbb{T}^\varkappa$ and $q \in (0,1]$. For any $t > 0$, we define the number $\Delta_t^q(f)(t)$ (its existence is allowed) with the following properties. For an arbitrary $\varepsilon > 0$, there exists a δ-neighbourhood $\mathcal{N}_t \subset \mathbb{T}$ of the point t such that

$$|[f(\sigma(t)) - f(s)]t^{1-q} - \Delta_t^q(f)(t)][\sigma(t) - s]| \leq \varepsilon|\sigma(t) - s| \text{ at all } s \in \mathcal{N}_t.$$

The expression $\Delta_t^q(f)(t)$ is called the *fractional-like derivative* of the function f of order $q \in (0,1]$ at the point $t \in \mathbb{T}^\varkappa$ *on the time scale*.

The *fractional-like exponential function* $E_f(r,0)$ on the time scale, for the function $f \in \mathcal{R}^q$, is given by the formula

$$E_f(r,0) = \exp\left(\int_0^r \xi_{\mu(s)}(f(s)s^{q-1})\Delta s\right)$$

for all $(0,r) \in \mathbb{T}$, where ξ_k is the *cylindrical transformation* $\mathbb{C}_k \to \mathbb{Z}_k$, where $\mathbb{C}_k = \{z \in \mathbb{C} : z \neq -\frac{1}{k}, k > 0\}$ and $\mathbb{Z}_k = \{Z \in \mathbb{C} : -\frac{\pi}{k} < I_m(z) < \frac{\pi}{k}\}$.

$\mathcal{R}_+^q$ is the set of positive q-regressive functions $\mathcal{R}^q = \{f : \mathbb{T} \to \mathbb{R} : 1 + \mu(t)f(t)t^{q-1} \neq 0\}$ for all $t \in \mathbb{T}^\varkappa$.

An operation Θ_q is defined for the function $f \in \mathcal{R}^q$ as

$$(\Theta_q f)(t) = -\frac{f(t)}{1 + \mu(t)f(t)t^{q-1}} \quad \forall t \in \mathbb{T}^\varkappa.$$

Next, we consider the function $f : \mathbb{T} \to \mathbb{R}$ and its fractional-like derivative on the time scale at the point $t \in \mathbb{T}^\varkappa$.

6.2.3　Fractional-like derivative and its properties

Here, we consider the function $f : \mathbb{T} \to \mathbb{R}$ and its fractional-like derivative on the time scale at the point $t \in \mathbb{T}^{\varkappa}$.

Some properties of the fractional-like derivative of the function f are stated as follows (cf. [168]).

Theorem 6.1. *Let $f : \mathbb{T} \to \mathbb{R}$ and $t \in \mathbb{T}^{\varkappa}$. Then, the following statements are true for the function f:*

(1) *If f is a fractionally similar differentiable function of order $q \in (0, 1]$ at the point $t > 0$, then f is continuous at the point t.*

(2) *If the function f is continuous at the point t and t is a scattered function, then f is fractionally differentiable of order q at the point $t \in \mathbb{T}^{\varkappa}$ and*

$$\Delta_t^q(f)(t) = \frac{f(\sigma(t)) - f(t)}{\mu(t)} t^{1-q}.$$

(3) *If f is a fractionally similar differentiable function of order q at the point $t \in \mathbb{T}^{\varkappa}$, then*

$$f(\sigma(t)) = f(t) + (\mu(t)) t^{q-1} \Delta_t^q(f)(t).$$

(4) *If $f, g : \mathbb{T} \to \mathbb{R}$ are fractionally similarly differentiable functions of order $q \in (0, 1]$, then*

$$\Delta_t^q(f + g)(t) = \Delta_t^q(f)(t) + \Delta_t^q(g)(t).$$

(5) *If $f : \mathbb{T} \to \mathbb{R}$ is a fractionally similar differentiable function, then for any $\lambda \in \mathbb{R}$,*

$$\Delta_t^q(\lambda f)(t) = \lambda \Delta_t^q(f)(t).$$

(6) *If the functions f and g are continuous, then the product $fg : \mathbb{T} \to \mathbb{R}$ is fractionally similarly differentiable according to the formula*

$$\left(\Delta_t^q(fg)\right)(t) = \left(\Delta_t^q(f)g\right)(t) + (f \circ \sigma(t))\left(\Delta_t^q(g)\right)(t)$$

$$= \left(\Delta_t^q(f)(g \circ \sigma(t))\right)(t) + \left(f\left(\Delta_t^q(g)\right)\right)(t) = (fg)^{\Delta}(t) t^{1-q},$$

where Δ denotes the ordinary Δ-derivative on the time scale.

(7) *If f is a continuous function, then $1/f$ is fractionally differentiable and*

$$\Delta_t^q\left(\frac{1}{f}\right)(t) = -\frac{\Delta_t^q(f)(t)}{f(f \circ \sigma(t))} = \left(\frac{1}{f}\right)^{\Delta}(t) t^{1-q}$$

for all points $t \in \mathbb{T}^k$ for which $f(t)f(\sigma(t)) \neq 0$.

(8) *If the functions f and g are continuous, then the function f/g is fractionally similarly differentiable and*

$$\Delta_t^q\left(\frac{f}{g}\right)(t) = \frac{\Delta_t^q(f)(t)g(t) - f(t)\left(\Delta_t^q(g)(t)\right)}{g(g(t) \circ \sigma(t))}) = \left(\frac{f}{g}\right)^\Delta (t)t^{1-q}$$

for all points $t \in \mathbb{T}^k$ for which $g(t)g(\sigma(t)) \neq 0$.

(9) *If the function $g : \mathbb{T} \to \mathbb{R}$ is continuous and fractionally differentiable of order $q \in (0,1]$ at the point $t \in \mathbb{T}^k$ and the function $f : \mathbb{R} \to \mathbb{R}$ is continuously differentiable, then there is a constant $c \in [t, \sigma(t)]$ for which*

$$\Delta_t^q(f(g))(t) = f(g(c))\Delta_t^q(g)(t).$$

The proofs of assertions (1)–(9) are derived according to the scheme of the proof of Theorems 6.1–6.3 from the monograph [20].

6.2.4 The fractional-like integral and its properties

The *fractional-like integral on the time scale* is introduced on the basis of two concepts: q-regularity and rd-continuity of the function $f : \mathbb{T} \in \mathbb{R}$.

Let $f : \mathbb{T} \in \mathbb{R}$ and q be a regular function. Then, the fractional-like integral of the function f for $q \in (0,1]$ is determined using the formula (see [17] and the bibliography therein)

$$I_q f(t) = \int f(t)\Delta^q t = \int f(t)t^{q-1}\Delta t.$$

Let us recall some *properties of the fractional-like integral* on the time scale.

Theorem 6.2. *Let $q \in (0,1]$, and let f,g be rd-continuous functions, the constant $\lambda \in \mathbb{R}$, and $a,b \in \mathbb{T}$. Then:*

(1) *The relation*

$$\int_a^b f(t)\Delta^q t = I_q(b) - I_q(a)$$

holds.

(2) *The relation*

$$\int_a^b [f(t) + g(t)]\Delta^q t = \int_a^b f(t)\Delta^q t + \int_a^b g(t)\Delta^q t$$

is true for the sum of two functions.

(3) *If $f(t) > 0$ for all $t \in [a, b]$, then $\int_a^b f(t) \Delta^q t \geq 0$.*

(4) *The equality*

$$\left(\Delta_t^q I_q f(t) \right)(t) = f(t) \quad \text{holds at all} \quad t \in \mathbb{T}.$$

(5) *If $\Delta_t^q(f)(t)$ is an rd-continuous function, then*

$$I_q \left(\Delta_t^q(f) \right)(t) = f(t) - f(a).$$

(6) *If $\Delta_t^q(f)(t) \leq 0$ on $[a, b]$, then the function $f(t)$ is decreasing on $[a, b]$.*

The proofs of assertions (1)–(6) are given in the paper [168].

6.3 Statement of the Problem

The following system of equations of perturbed motion on a time scale with a fractional-like derivative of the state vector is considered:

$$\Delta_t^q(x)(t) = f(t, x(t)), \tag{6.1}$$

$$x(0) = x_0, \tag{6.2}$$

where $\Delta_t^q(x)(t)$ is the fractional-like derivative of the vector $x \in \mathbb{R}^n$ on the time scale $\mathbb{T}$, $F(t) = f(t, x(t)) \in C_{rd}(\mathbb{T} \times \mathbb{R}^n, \mathbb{R}^n), t \in \mathbb{T}, q \in (0, 1]$.

Further, it is assumed that the initial problem (6.1)–(6.2) has a solution $x(t, 0, x_0) = x(t)$, with the initial values $(0, x_0) \in (\mathbb{R}_+ \times \mathbb{R}^n)$ for all $t \geq 0$.

The fractional Δ is a derivative of the complex function $\Delta_t^q(f \circ g)(t)$ on the time scale; it is used further in this chapter.

We know that

$$\left[f(g(t)) \right]^\Delta = \left(\int_0^1 f'(g(t) + h\mu(t) g^\Delta(t)) dh \right) g^\Delta(t)$$

(see [3, 9, 115] and the bibliography therein).

If $q = 1$ and the function $V(x(t)) = x^T P(t) x$ is Δ-differential on the time scale $\mathbb{T}$, then

$$\left[V(x(t)) \right]^\Delta = \left(x^T P(t) x \right)^\Delta = (x^T P(t)) x^\Delta + (x^T P(t))^\Delta x(\sigma(t)).$$

Together with the system of equations (6.1), consider the function $V(t, x) \in C_{rd}(\mathbb{R}_+ \times \mathbb{R}^n, \mathbb{R}_+)$ such that $V(t, 0) = 0$ for all $t \geq 0$ and $x \in B_r = (x \in \mathbb{R}^n : \|x\| < r), r > 0$. If the function $V = V(x)$ is autonomous, then at the solutions of the system (6.1), the function $(\Delta_t^q V)(t)$ will depend on t due to the non-constant function of the granularity of the time scale $\mathbb{T}$.

Let the function $V(t, x)$ be fractionally differentiable and

$$\Delta_t^q V(t, x)(t) = \left[V(t, x(t)) \right]^\Delta \Delta_t^q(x)(t). \tag{6.3}$$

Remark 6.1. Since $V(t, x(t)) = v(t)$ on the solutions of system (6.1), according to statement (2) of Theorem 6.1, we obtain

$$\Delta_t^q(v)(t) = \frac{v(x(\sigma(t))) - v(x(t))}{\mu(t)} t^{1-q}$$

if $v(t)$ is continuous at any point $t \in \mathbb{T}^\varkappa$ in a scattered case.

A qualitative analysis of the properties of the solutions of the fractional-like system of equations (6.1) is of interest, as well as the problem of stability and boundedness of solutions, in particular.

6.4 Estimates of Lyapunov Functions on Solutions of System (6.1)

As in the theory of stability on the time scale (see [90]), in the theory of stability of solutions of fractional-like equations on the time scale, the following *Lyapunov relation* is of key importance:

$$V(t, x(t)) = V(0, x_0) + I_q \Delta_t^q(V(t, x(t))(t),$$

where I_q is a fractional-like integral on the scale $\mathbb{T}$. Let us show that the following statement is true.

Lemma 6.1. *Let for system (6.1), there exist a Lyapunov function* $V(t, x) \in C_{rd}^q(\mathbb{R}_+ \times \mathbb{R}^n, \mathbb{R}_+)$*, the function* $f \in C_{rd}(\mathbb{R}_+)$ *and the exponential function* E_p *be defined for the function* $p(t) \in \mathcal{R}^q(t, 0)$*. Then, if for all* $t \in \mathbb{T}^k$*, the inequality*

$$\Delta_t^q V(t, x)(t) \leq p(t) V(t, x) + f(t), \tag{6.4}$$

holds, then

$$V(t, x(t)) \leq V(0, x_0) E_p(t, 0) + \int_0^t E_p(t, \sigma(s)) f(s) \Delta^q s \tag{6.5}$$

for all $t \in \mathbb{T}$*.*

Proof. Given that $E_p(t, 0) = \frac{1}{E_p(0,t)} = E_{\ominus qp}(t, 0)$, and according to statement (6) of Theorem 6.1, we obtain

$$\Delta_t^q \left(V(t, x) E_{\ominus qp}(t, 0) \right)(t)$$

$$= \Delta_t^q (V(t, x))(t) E_{\ominus qp}(\sigma(t), 0) + V(t, x) \Delta_t^q \left(E_{\ominus qp}(t, 0) \right). \tag{6.6}$$

It follows from relation (6.6) that

$$\Delta_t^q\left(V(t,x)E_{\ominus_{qp}}(t,0)\right)(t) = \Delta_t^q(V(t,x))(t) - p(t)V(t,x)E_{\ominus_{qp}}(\sigma(t),0). \quad (6.7)$$

According to statement (5) of Theorem 6.2, from (6.7), we obtain the Lyapunov relation for the function $V(t,x)$ in the form

$$V(t,x)E_{\ominus_{qp}}(t,0) - V(0,x_0)E_{\ominus_{qp}}(v,0)$$

$$= \int_0^t \left(\Delta_t^q(Vt,x)(s) - p(s)(V(s,x(s)))\right)E_{\ominus_{qp}}(\sigma(s),0)\Delta^q s. \quad (6.8)$$

It follows from (6.8) that

$$V(t,x(t)) \leq V(0,x_0)E_p(t,v) + \int_0^t \frac{E_{\ominus_{qp}}(\sigma(s),v)}{E_{\ominus_{qp}}(t,0)}f(s)\Delta^q s$$

or, finally in kind,

$$V(t,x(t)) \leq V(0,x_0)E_p(t,0) + \int_0^t f(s)E_p(t,\sigma(s))\Delta^q s$$

for all $t \in \mathbb{T}$. Thus, Lemma 6.1 is proved. $\square$

Lemma 6.2. *Let the following conditions be satisfied for system* (6.1):

(1) *there exists a Lyapunov function* $V(t,x) \in C_{rd}^q(\mathbb{R}_+ \times \mathbb{R}^n, \mathbb{R}_+)$;
(2) *there exists a function* $f_1 \in C_{rd}$, *and for a function* $p_1(t) \in R_+^q$, $p_1(t) \geq 0$ *for all* $t \in T^k$ *is a defined exponential function* $E_{p_1}(t,\sigma(t))$;
(3) *for all* $t \in \mathbb{T}$, *the estimate*

$$V(t,x(t)) \leq f_1(t) + \int_0^t V(s,x(s))p_1(s)\Delta^q s \quad (6.9)$$

is valid. Then,

$$V(t,x(t)) \leq f_1(t) + \int_0^t E_{p_1}(t,\sigma(s))f_1(s)\Delta^q s \quad (6.10)$$

for all $t \in \mathbb{T}$.

Proof.   Let us introduce the notation

$$W(t) = \int_0^t V(s,x(s))p_1(s)\Delta^q s \quad (6.11)$$

for all $t \in \mathbb{T}$. It is obvious that $W(0) = 0$ and

$$V(t,x(t)) \leq f_1(t) + W(t). \quad (6.12)$$

By performing a fractional-like differentiation of relation (6.11), we obtain

$$\Delta_t^q W(t) = V(t, x(t))p_1(t) \leq f_1(t)p_1(t) + p_1(t)W(t). \tag{6.13}$$

It follows from (6.13) that

$$W(t) \leq \int_0^t E_{p_1}(t, \sigma(s))p_1(s)\Delta^q s \tag{6.14}$$

for all $t \in \mathbb{T}$.

In view of estimate (6.12), from inequality (6.14), we obtain estimate (6.10).

Thus, Lemma 6.2 is proved. $\qquad\square$

Lemma 6.3. *Let the following conditions for system* (6.1) *hold:*

(1) *there exists a Lyapunov function* $V(t, x) \in C_{rd}^q(\mathbb{R}_+ \times \mathbb{R}^n, \mathbb{R}_+)$ *for all* $(0, x) \in B_r$;

(2) *there exists a function* $p_2(t) \in C_{rd}$, $p_2(t) \in \mathcal{R}^q$ *for which*

$$\Delta_t^q(V(t, x))(t) \leq p_2(t)V(t, x) \tag{6.15}$$

for all $(t, x) \in [0, \tau] \times B_r$.

Then,

$$V(t, x(t)) \leq V(0, x_0)E_{p_2}(t, 0) \tag{6.16}$$

for all $t \in [0, \tau] \in \mathbb{T}$.

Proof. From estimate (6.15), we get

$$V(t, x(t)) \leq V(0, x_0) + \int_0^t V(s, x(s))p_2(s)\Delta^q s \tag{6.17}$$

for all $t \in [0, \tau] \in \mathbb{T}$.

Applying Lemma 6.2 to estimate (6.17), for $f(t) = V(0, x_0)$, we obtain

$$V(t, x(t)) \leq V(0, x_0) + V(0, x_0) \int_0^t E_{p_2}(t, \sigma(s))p_2(s)\Delta^q s. \tag{6.18}$$

According to Theorem 3.12 from the paper [17], estimate (6.18) is reduced to the form

$$V(t, x(t)) \leq V(0, x_0)(1 + E_{p_2}(t, 0) - E_{p_2}(t, t)) = V(0, x_0)E_{p_2}(t, 0)$$

for all $t \in [0, \tau] \in \mathbb{T}$.

Thus, Lemma 6.3 is proved. $\qquad\square$

6.5   Applications

Estimates (6.5), (6.10), and (6.16) of the function $V(t, x)$ on the solutions of
the system (6.1) allow us to carry out a qualitative analysis of the solutions
of the system (6.1).

6.5.1   Boundedness of the solutions

In this section, sufficient conditions for the boundedness of solutions of
system (6.1) based on the estimation of the Lyapunov function (6.5) are
presented.

Let the vector function $F(t) = f(t, x(t))$ in system (6.1) be such that
$F(t) \neq 0$ for $x(t) = 0$ and for all $t \in \mathbb{T}$.

Definition 6.3 (cf. [114]). The solution $x(t)$ of the initial problem (6.1)–
(6.2), $(0, x_0) \in \mathbb{T} \times \mathbb{R}^n$, is *uniformly bounded* if for any $q \in (0, 1]$, there
exists a constant $\beta > 0$ such that $\|x(t)\| \leq \beta$ for all $t \in [0, \infty)$, where the
value of β may depend on each solution.

The following statement holds.

Theorem 6.3. *Let* $B_r \subset \mathbb{R}^n$ *and there exist a function* $V(t, x)$, $V \in$
$C_{rd}(\mathbb{T} \times B_r \to [0, \infty))$ *such that for all* $(t, x) \in [0, \infty) \times B_r$, *all the conditions
of Lemma* 6.1 *are fulfilled. In addition:*

(1) *there exists* $c_1 > 0$ *such that* $c_1\|x\|^2 \leq V(t, x)$, $V(t, x) \to \infty$ *as* $\|x\| \to$
∞;
(2) *there is a non-decreasing function* $\phi(\|x\|)$ *such that* $V(t, x) \leq \phi(\|x\|)$;
(3) *for all* $t \in [0, \infty)$ *for some* $\beta > 0$, *the following estimate is valid:*

$$E_p(t, 0) + \frac{1}{\phi(\|x_0\|)} \int_0^t E_p(t, \sigma(s)) f(s) \Delta^q s \leq \frac{\beta c_1}{\phi(\|x_0\|)}.$$

Then, all solutions $x(t)$ *of the initial problem* (6.1)–(6.2) *remain in
the domain* B_r *and are considered uniformly bounded.*

Proof. Let $x(t)$ be some solution of the initial problem (6.1)–(6.2), which
remains in the domain B_r for all $t \geq 0$. When conditions (1) and (2) of
Theorem 6.3 are fulfilled, from inequality (6.5), we obtain the estimate

$$c_1\|x(t)\|^2 \leq \phi(\|x_0\|) \left(E_p(t, \upsilon) + \frac{1}{\phi(\|x_0\|)} \int_0^t E_p(t, \upsilon(s)) f(s) \Delta^q s \right). \quad (6.19)$$

From the estimate in (6.19), under condition (3) of Theorem 6.3, we find that

$$\|x(t)\| < \beta^{1/2}(\|x_0\|), \tag{6.20}$$

where β depends only on x_0. Consequently, all solutions of the initial problem (6.1)–(6.2) that remain in the domain B_r are considered uniformly bounded. Thus, Theorem 6.3 is proved. □

6.5.2 Motion with a given installation time

Let us determine one of the qualitative properties of the motion of system (2.1) as follows.

Definition 6.4. For a given function $V \in C_{rd}(\mathbb{T} \times B_r, [0, \infty))$ and a constant $a > 0$, the *motion* of system (6.1) is *established* at the moment $\tau \in [0, \infty)_{\mathbb{T}}$ on the set $V(t, x) \leq a$ if there exists $\tau \in \mathbb{T}$ such that for any solution $x(t)$, the condition $V(t, x(t)) \leq a$ is fulfilled for all $t \geq \tau$ and for any $x_0 \in \mathbb{R}^n$ for which $V(0, x_0) < \infty$.

The following statement holds.

Theorem 6.4. *Let $x_0 \in \mathbb{R}^n$, there exist a function $V \in C_{rd}(\mathbb{T} \times B_r, [0, \infty))$, for $\|x\| \to \infty$, all the conditions of Lemma 6.2 be fulfilled. In addition:*

(1) *there exists a constant $\lambda > 0$ such that $\lambda\|x\|^p \leq V(t, x)$ for all $(t, x) \in \mathbb{T} \times \mathbb{R}^n$ and $p > 1$;*

(2) *for some $a > 0$,*

$$\tau = \sup\left(t \in \mathbb{T} : 1 + \frac{1}{f_1(t)} \int_0^t E_{p_1}(t, \sigma(s)) f_1(s) \Delta^q s \leq \frac{a}{f_1(t)}\right)$$
$$< +\infty.$$

Then, the solution $x(t)$ of system (6.1) is established on the set $V(t, x) < a$ for all $t \in [0, \tau)$.

Proof. Let $x(t)$ be the solution of system (6.1) at $x(0) = x_0 \in \mathbb{R}^n$, for which $V(0, x_0) < \infty$. It follows from Lemma 6.2 and condition (2) of Theorem 6.4 that

$$V(t, x(t)) \leq f_1(t)\left(1 + \frac{1}{f_1(t)} \int_0^t E_{p_1}(t, \sigma(s)) f_1(s) \Delta^q s\right). \tag{6.21}$$

From the estimate in (6.21), under condition (2) of Theorem 6.4, we have the estimate

$$V(\tau, x(\tau)) \leq a.$$

Consequently, the solution $x(t)$ remains in the set $V(t, x) \leq a$ for all $t \in [0, \tau)$. If condition (1) of Theorem 6.4 is satisfied, then

$$\|x(t)\|^p \leq \frac{f_1(t)}{\lambda} \left(1 + \frac{1}{f_1(t)} \int_0^t E_{p_1}(t, \sigma(s)) f_1(s) \Delta^q s \right),$$

and when the following inequality is fulfilled for all $t \in [0, \tau)$,

$$1 + \frac{1}{f_1(t)} \int_0^t E_{p_1}(t, \sigma(s)) f_1(s) \Delta^q s \leq \frac{\lambda}{f_1(t)} V(0, x_0),$$

we get the estimate

$$\|x(t)\| \leq (V(0, x_0))^{1/p} = R \tag{6.22}$$

for some $R > 0$, where R depends on x_0. Consequently, the motion of system (6.1) is limited. $\qquad\square$

6.5.3   Practical stability

In this section, the motion of system (6.1) is considered, with some predetermined values $0 < \alpha < \beta$ limiting the initial values of x_0 and the solution $x(t)$ for all $t \in \mathbb{T}$.

Definition 6.5. System (6.1) is said to be *practically stable* if, given (α, A) with $0 < \alpha < A$, the condition $\|x_0\| \leq \alpha$ implies $\|x(t)\| < \beta$ for all $t \in \mathbb{T}$.

Let us indicate the sufficient conditions for this kind of stability of solutions $x(t)$ based on the estimate of the Lyapunov function from Lemma 6.3.

Theorem 6.5. *Let the following conditions for system* (6.1) *hold:*

(1) *there is a function* $V(t, x) : [0, \infty) \times B_\beta \to [0, \infty)$, *and the functions* $a \in KR$-*class and* $b \in K$-*class such that* $a(\|x\|) \leq V(t, x)$ *and* $V(t, x) \leq b(\|x\|)$;
(2) *for all* $t \in \mathbb{T}$, *all the conditions of Lemma 6.3 are fulfilled;*
(3) *at* $\|x_0\| < \alpha$, *the following inequality holds:*

$$a^{-1}(E_{p_2}(t, 0)) < \frac{\beta}{a^{-1}(b(\alpha))} \quad \forall t \in \mathbb{T}.$$

Then, the solution $x(t)$ *of system* (6.1) *is practically stable.*

Proof.   Let at $\|x_0\| < \alpha$, the solution of system (6.1) be defined for all $t \geq 0$. When condition (1) and (2) of Lemma 6.3 is fulfilled, we have for all $t \in \mathbb{T}$,

$$a(\|x(t)\|) \leq V(t, x(t)) \leq b(\|x_0\|) E_{p_2}(t, 0).$$

Hence,
$$\|x(t)\| \le a^{-1}(b(\alpha)E_{p_2}(t,0)),$$
and under condition (3) of Theorem 6.5, we obtain $\|x(t)\| < \beta$ for all $t \in \mathbb{T}$.
Thus, Theorem 6.5 is proved. $\qquad\square$

6.5.4 Lyapunov stability

Let in the system of equations (6.1), the vector function $f(t,x) = 0$ for $x = 0$ and for all $t \in \mathbb{T}$.

Definition 6.6. The trivial *solution* $x = 0$ of system (6.1) is said to be *equistable* if for each $\varepsilon > 0$, $t_0 \in \mathbb{T}$, there exists a positive function $\delta = \delta(t_0, \varepsilon)$ which is continuous in t_0 for each ε such that $\|x_0\| < \delta$ implies $\|x(t)\| < \varepsilon$ for all $t \in \mathbb{T}$.

Let us show that the following statement is true.

Theorem 6.6. *Let the following conditions hold for system* (6.1)*:*

(1) *there exists a function $V(t,x) \in C_{rd}(\mathbb{T} \times B_r, \mathbb{R}_+)$, the functions $a, b \in K$-class are such that*
$$a(\|x\|) \le V(t,x) \quad \text{and} \quad V(t,x) \le b(\|x\|) \quad \text{on} \quad [0,\infty) \times B_r;$$

(2) *all the conditions of Lemma 6.3 are fulfilled for the function $V(t,x)$ specified in condition (1) of this theorem;*

(3) *for all $t \in \mathbb{T}$, the inequality*
$$E_{p_2}(t,0) < \frac{a(\varepsilon)}{b(\delta)} \tag{6.23}$$

holds.

Then, the solution $x = 0$ of system (6.1) *is stable.*

Proof. When conditions (1) and (2) of Theorem 6.6 are fulfilled, from Lemma 6.3, we get the estimate
$$V(t,x(t)) \le b(\|x_0\|)E_{p_2}(t,0) < b(\delta)E_{p_2}(t,0) \tag{6.24}$$
for the initial conditions $x_0 : \|x_0\| < \delta$. Further, when inequalities (6.24) are satisfied, we have
$$a(\|x(t)\|) \le V(t,x(t)) < a(\varepsilon)$$
for all $t \in \mathbb{T}$. From here, it follows that
$$\|x(t)\| \le a^{-1}(V(t,x(t))) < a^{-1}(a(\varepsilon)) = \varepsilon$$
for all $t \in \mathbb{T}$ for any $\varepsilon > 0$. Therefore, Theorem 6.6 is proved. $\qquad\square$

6.6 Comments and Bibliography

This chapter examines a new class of dynamic equations with a fraction-like derivative of the state vector. Sufficient conditions for various types of stability of solutions are obtained on the basis of integral estimates of Lyapunov functions on solutions of the equations under consideration.

Section 6.2 contains preliminary results adapted from Bohner and Peterson [20], Keller [58], Khalil *et al.* [60], Potzsche [115], and Younus *et al.* [168]. In the paper by Wang *et al.* [170], fractional-like analysis in Sobolev space on a time scale is introduced (see also Rahmat [141]). The proposed approach is used to solve differential equations with fractional-like derivative of the state vector on time scales.

Sections 6.3–6.5 are based on the paper by Martynyuk [92].

The methods for analysing the stability of solutions to dynamic equations with the Δ-derivative of the state vector are summarised in the monograph by Martynyuk [90], in which the direct Lyapunov method is used based on scalar-, vector-, and matrix-valued auxiliary functions. The book by Anderson and Georgiev [7] summarises the most recent advances in this field. The authors have significantly expanded the known results and created a comprehensive theory developed specifically for this book.

Chapter 7

Equilibrium Stability under Nuclear Confrontation

"There lies before us, if we choose, continual progress in happiness, knowledge, and wisdom. Shall we, instead, choose death, because we cannot forget our quarrels? We appeal as human beings to human beings: Remember your humanity, and forget the rest. If you can do so, the way lies open to a new Paradise; if you cannot, there lies before you the risk of universal death."[1]

7.1 Introduction

This chapter proposes and analyses mathematical models of confrontation between two and n countries, including countries with nuclear weapons. The proposed models are based on a generalisation of the well-known Richardson's mathematical model of the arms race. Namely, the factor of hostility is filled with expanded content, including public opinion and the armed forces of the opposing countries. Qualitative analysis of confrontation models is carried out using the method of Lyapunov functions and by applying nonlinear integral inequalities. As a result of the analysis, the conditions for the stability of the equilibrium state of the opposing countries are established, and the influence of hostility on the decrease (increase) in the norm of the armament vector of n countries involved in alliances is shown. In the final section, for the general mathematical model of opposition of n countries, estimates of the deviation of the armament vector from the equilibrium ray are given.

---

[1] *See the Russell–Einstein Manifesto, available on the internet.*

This chapter is organised according to the following plan.

Section 7.2 presents a quasi-linear non-autonomous mathematical model of a confrontation between two countries. In this model, nonlinear functions of hostility are assumed to depend on the level of public opinion about the confrontation between the two countries and their weapons.

In Section 7.3, the stability of the mathematical model under zero hostility and under structural perturbations of the linear approximation is investigated.

Section 7.4 examines the situation when the linear approximation in the mathematical model is neutrally stable, starting from a certain point in time. In addition, here, we establish the conditions for the limited level of armaments of the opposing countries.

Section 7.5 examines the problem of stability (instability) of equilibrium in the generalised mathematical model of Richardson under various assumptions about the level of hostility between countries.

In Section 7.6, for the mathematical model with an autonomous linear approximation, the stability conditions of the zero equilibrium state are established using the Lyapunov function method. Under various assumptions on the level of hostility compared to the threat levels and the level of defence expenditures, formulas for changing the level of armaments of the two countries are given. At the boundary level of hostility between the two countries, the boundary of the variation of the mobile equilibrium state is established.

Section 7.7 discusses the problem of equilibrium stability in the model of confrontation between n countries involved in alliances. Conditions for the decrease (increase) in the level of armament depending on the level of hostility between countries are obtained.

Section 7.8 considers a general non-autonomous nonlinear mathematical model of confrontation between n countries. For this model, the stability conditions for the deviation of the armament vector from the arms race equilibrium ray are established using the method of nonlinear integral inequalities.

The final section contains comments and bibliography for the chapter.

7.2 Generalised Richardson Arms Race Model

Based on the Richardson *mathematical model* [145] of the arms race, we describe a model of confrontation between two countries N_1 and N_2 as a non-autonomous quasi-linear system of differential equations:

$$\frac{dx}{dt} = k(t)y - a(t)x + g(t,x,y), \quad \frac{dy}{dt} = l(t)x - b(t)y + h(t,x,y), \quad (7.1)$$

where $t \in \mathbb{R}_\tau, \tau$ is a finite number or symbol $+\infty$, $x,y \in \mathbb{R}_+$ are the measures of the level of armament of countries N_1 and N_2, the parameters $k(t) > 0$ and $l(t) > 0$ characterise the *"threat" between countries*, and the parameters $a(t) > 0$ and $b(t) > 0$ characterise the *"expenses" of each country* to maintain its level of armament for all $t \in \mathbb{R}_\tau$. The parameters $k(t) > 0$, $l(t) > 0$ and $a(t) > 0$, $b(t) > 0$ are bounded for any $t \in \mathbb{R}_\tau$.

The functions $g(t,x,y)$ and $h(t,x,y)$ characterise the *"hostility" factor between countries* N_1 and N_2, which depends on the level of public opinion about the confrontation between the countries and their armament.

Remark 7.1. If countries N_1 and N_2 possess nuclear weapons, then the parameters $k(t)$ and $l(t)$ can take large values for any $t \in R_\tau$ while remaining bounded.

It is assumed that the hostility between countries N_1 and N_2 does not exceed some threshold level $b = (b_1, b_2)^T$ on any finite time interval for any levels of armament of the opposing countries.

Thus, for the components of the hostility vector function $b(t,x,y) = (g(t,x,y), h(t,x,y))^T$, there exist constants $\bar{b}_1 > 0$ and $\bar{b}_2 > 0$ such that

$$|g(t,x,y)| \leq \bar{b}_1, \quad |h(t,x,y)| \leq \bar{b}_2 \qquad (7.2)$$

for all $t \in \mathbb{R}_\tau$ and $x,y \in \mathbb{R}_+$.

Remark 7.2. The maximum values of permissible hostility $\bar{b}_1$ and $\bar{b}_2$ between the countries N_1 and N_2 should be under the control of the civil society of these countries and/or international organisations, given that modern means of influencing the human psyche have reached the level of criminality.

7.3 Stability Conditions for Zero Hostility in Time

Along with the system of differential equations (7.1), we consider the system of equations

$$\frac{dx}{dt} = k(t)y - a(t)x, \quad \frac{dy}{dt} = l(t)x - b(t)y \qquad (7.3)$$

with zero hostility level $g(t,x,y) = 0$ and $h(t,x,y) = 0$ for any $t \in \mathbb{R}_\tau$.

Proposition 7.1. *If in the model* (7.3), *the parameters* $k(t), a(t), l(t),$ *and* $b(t)$ *are such that*

$$(1) \quad -a(t) \leq 0 \quad \textit{for all} \quad t \in \mathbb{R}_\tau, \tag{7.4}$$

$$(2) \quad a(t)b(t) - (k(t) + l(t))^2 \geq 0 \quad \textit{for all} \quad t \in \mathbb{R}_\tau, \tag{7.5}$$

then the equilibrium state $(x = 0, y = 0)^T$ *is stable.*

Proof. For a definitely positive Lyapunov function $V(x, y) = \frac{1}{2}(x^2 + y^2)$, it is easy to show that under conditions (1) and (2) of Proposition 7.1, its total derivative by virtue of the system (7.3)

$$\frac{d}{dt}V(x, y) = \frac{dx}{dt}x + \frac{dy}{dt}y = x(k(t)y - a(t)x) + y(l(t)x - b(t)y) \tag{7.6}$$

is not positive. In this case, according to Lyapunov's Theorem 1 (see [76, pp. 61–62]), the equilibrium state $(x = 0, y = 0)^T$ of the system (7.3) is asymptotically stable. $\qquad\square$

Proposition 7.2. *If in the model* (7.3), *the parameters* $k(t), a(t), l(t),$ *and* $b(t)$ *are such that there exist positive numbers* α *and* β *and the conditions*

$$(1) \quad -a(t) < -\alpha \quad \textit{for all} \quad t \in \mathbb{R}_\tau, \tag{7.7}$$

$$(2) \quad a(t)b(t) - (k(t) + l(t))^2 > \beta \quad \textit{for all} \quad t \in \mathbb{R}_\tau, \tag{7.8}$$

are satisfied, then the equilibrium state $(x = 0, y = 0)^T$ *is asymptotically stable.*

Proof. Under conditions (1) and (2) of Proposition 7.2, the total derivative of the function $V(x, y) = \frac{1}{2}(x^2 + y^2)$ is negative definite due to the system (7.3). According to Remark 7.2 to Lyapunov's Theorem 1 (see [76, p. 64]), the equilibrium state $(x = 0, y = 0)^T$ of the system (7.3) is asymptotically stable.

Next, we rewrite the system (7.3) in vector-matrix form:

$$\frac{dw}{dt} = A(t)w, \quad w(t_0) = w_0, \tag{7.9}$$

where $w = (x, y)^T$ and $A(t)$ is a bounded 2×2 matrix of weapons, and we consider the problem of the stability of the equilibrium state $(x = 0, y = 0)^T$ under parametric perturbations of the matrix of weapons.

Let, at the moment $t = t^* \in \mathbb{R}_\tau$, the matrix of weapons $A(t^*)$ has eigenvalues λ satisfying the condition $Re\lambda \leq -\alpha$, i.e., the equilibrium state $(x = 0, y = 0)^T$ of the system (7.9) with the armament matrix $A(t^*)$ is asymptotically stable.

We rewrite the system (7.9) as

$$\frac{dw}{dt} = A(t^*)w + \Delta A(t)w, \quad w(t^*) = w^*, \tag{7.10}$$

where the notation $\Delta A(t) = A(t) - A(t^*)$ is adopted.

We associate the Lyapunov function $V(w) = w^T w$ with the system (7.9) and assume that the following conditions are satisfied:

(1) $w^T(A^T(t^*) + A(t^*))w \le -\lambda_M w^T w$ for all $t \in \mathbb{R}_\tau$, (7.11)

(2) $w^T(\Delta A^T(t) + \Delta A(t))w \le \mu_M(t)w^T w$ for all $t \in \mathbb{R}_\tau$, (7.12)

where λ_M is the maximum eigenvalue of the matrix $A^T(t^*) + A(t^*)$ and $\mu_M(t)$ is the maximum eigenvalue of the matrix $\Delta A^T(t) + \Delta A(t)$ for all $t \in \mathbb{R}_\tau$. □

Proposition 7.3. *If in the model (7.10), the parameters $k(t), a(t), l(t)$, and $b(t)$ are such that the conditions (7.11) and (7.12) are satisfied and if for any $\varepsilon > 0$, there exists $\delta = \delta(t^*, \varepsilon) > 0$ such that*

$$\exp\left(-\lambda_M(t - t^*) + \int_{t^*}^t \mu_M(s)ds\right) < \frac{\varepsilon}{\delta} \quad \text{for all} \quad t \in \mathbb{R}_\tau,$$

then the equilibrium state $w = 0$ of the system (7.10) is stable.

Proof. For the total derivative of a positive definite function $V(w) = w^T w$, under the conditions (7.11) and (7.12), we have the estimate

$$\frac{d}{dt}V(w(t)) \le (-\lambda_M + \mu_M(t))V(w(t)) \quad \text{for all} \quad t \in \mathbb{R}_\tau. \tag{7.13}$$

It follows from the estimate (7.13) that

$$V(w(t)) \le V(w^*)\exp\left(\int_{t^*}^t (-\lambda_M + \mu_M(s))ds\right) \quad \text{for all} \quad t \in \mathbb{R}_\tau.$$

From here, when the conditions of Proposition 7.3 are fulfilled, the stability of the equilibrium state $w = 0$ of the system (7.10) follows under *parametric perturbations of the armament matrix $A(t^*)$.* □

7.4 Stability in a Neutral Linear Approximation of the Model

Let us consider a situation in which, at the moment $t_r \in \mathbb{R}_\tau$, the linear approximation of the system (7.1) is neutrally stable. This is the case if

$$k(t_r)y - a(t_r)x = 0, \quad l(t_r)x - b(t_r)y = 0. \tag{7.14}$$

In this case, the mathematical model of confrontation between two countries (7.1) becomes

$$\frac{dx}{dt} = g(t,x,y), \quad x(t_r) = x_0^r; \quad \frac{dy}{dt} = h(t,x,y), \quad y(t_r) = y_0^r \qquad (7.15)$$

for all $t \geq t_r$. Assume that the hostility functions $g(t,x,y)$ and $h(t,x,y)$ are essentially nonlinear (i.e., they do not contain a linear approximation).

We associate the constant-sign bounded Lyapunov function $V(x,y) = xy, x > 0, y > 0$ with the system of equations (7.15) and show that the following statement holds.

Proposition 7.4. *Let $g(t,0,0) = 0$ and $h(t,0,0) = 0$ for all $t \geq t_r$ in the mathematical model of confrontation* (7.15)*. If the inequality*

$$g(t,x,y)y + h(t,x,y)x \geq \theta_0(t)xy \ \text{with} \ \theta_0(t) > 0, \ \text{and} \ \int_0^\infty \theta_0(s)ds = \infty \tag{7.16}$$

is satisfied for all $t \geq t_r$, then the equilibrium state $x = y = 0$ of the system (7.15) *is unstable.*

Proof. When the inequality (7.16) is satisfied, the total derivative of the function $V(x,y) = xy$ is infinitely increasing. Hence, by virtue of the Lyapunov instability theorem (see [76, pp. 65–66]), the instability of the equilibrium state $x = y = 0$ of the system (7.15) follows.

Note that if in inequality (7.16) the function $\theta_0(t) = 0$, then for any arbitrarily small values of the functions $g(t,x,y) > 0, h(t,x,y) > 0$ for all $t \geq t_r$ with neutral stability of the linear approximation in the mathematical model (7.1) the equilibrium state $x = y = 0$ can be stable.

Next, we consider the dynamics in the mathematical model (7.15) in the range of values $(t,x,y) \in \mathbb{R}_\tau \times \mathbb{R}_+ \times \mathbb{R}_+$ under the following assumptions.

There exist non-negative integrable functions $\psi_1(t)$ and $\psi_2(t)$ such that, in the range of values $(t,x,y) \in \mathbb{R}_\tau \times \mathbb{R}_+ \times \mathbb{R}_+$, the following estimates are satisfied:

(a) $\quad |g(t,x,y)y| \leq \psi_1(t)(xy)^p,$

(b) $\quad |h(t,x,y)x| \leq \psi_2(t)(xy)^q,$ $\qquad\qquad (7.17)$

(c) $\quad g(t,0,0) \neq 0, h(t,0,0) \neq 0,$

where $1 < p, 1 \leq q < \infty$. $\qquad\qquad\qquad\qquad\qquad\qquad\qquad\qquad\qquad \square$

We introduce the following definitions.

Definition 7.1. An *arms race* between two countries described by the system of equations (7.15) is:

(a) β-bounded if there exists a constant $\beta > 0$ such that $x(t, t_0, x_0^r)$ $y(t, t_0, y_0^r) < \beta$ for all $t \geq t_r$, where β may depend on each solution;
(b) equi-β-bounded if for any $\alpha > 0$ and $t_r \in \mathbb{R}_\tau$, there exists a constant $\bar{\beta}(t_0, \alpha) > 0$ such that if $x_0^r y_0^r < \alpha$, then $x(t, t_0, x_0^r)y(t, t_0, y_0^r) < \bar{\beta}(t_r, \alpha)$ for all $t \geq t_r$;
(c) uniformly β-bounded if the value $\bar{\beta}(t_r, \alpha)$ in Definition 7.1(b) does not depend on t_r.

For the Lyapunov function $V(x, y) = xy$, the following statement holds.

Lemma 7.1. *Assume that conditions (a)–(c) are satisfied and the inequality*

$$N(t, t_r) = (p + q - 2) \left[(V(x_0^r, y_0^r))^{p-1} \int_{t_r}^t \psi_1(s)ds + (V(x_0^r, y_0^r)^{q-1} \right.$$

$$\left. \times \int_{t_r}^t \psi_2(s)ds \right] < 1 \qquad (7.18)$$

holds for all $t \geq t_r$. Then, on the solutions of the system of equations (7.15), for the function $V(x, y)$, the estimate

$$V(x(t), y(t))) \leq V(x_0^r, y_0^r)\left(1 - N(t, t_r)\right)^{-\frac{1}{p+q-2}} \qquad (7.19)$$

is satisfied for all $t \in J$.

Proof. If conditions (a)–(c) are satisfied, it is easy to obtain the integral inequality for the function $V(x, y)$,

$$V(x(t), y(t)) \leq V(x_0^r, y_0^r) + \int_{t_r}^t (\psi_1(s)V^p(x(s), \ y(s))ds$$

$$+\psi_2(s)V^q(x(s), \ y(s)))ds, \quad t \geq t_r. \qquad (7.20)$$

The assertion of Lemma 7.1 is obtained by applying Theorem A.3.1 from [124, pp. 179–181] to the nonlinear integral inequality (7.20). □

Lemma 7.1 allows us to establish new conditions for the boundedness of the arms race between two opposing countries.

Proposition 7.5. *Assume that for the system (7.15), all conditions of Lemma 7.1 are satisfied and, in addition, for the initial levels of weapons x_0^r and y_0^r for any $\alpha > 0$, $x_0^r y_0^r < \alpha$, the following conditions are satisfied:*

(1) *$\bar{N}(t, t_r) < 1$ for all $t \geq t_r$;*
(2) *for a given value of $\bar{\beta} > 0$, the condition*

$$\left(1 - \bar{N}(t, t_r)\right)^{-\frac{1}{p+q-2}} < \frac{\bar{\beta}(\alpha)}{\alpha}$$

holds for all $t \geq t_r$, where

$$\bar{N}(t, t_r) = (p + q - 2)\left[(\alpha)^{p-1} \int_{t_r}^t \psi_1(s)ds + (\alpha)^{q-1}\right.$$

$$\left. \times \int_{t_r}^t \psi_2(s)ds\right]. \tag{7.21}$$

Then, the arms race between two opposing countries is uniformly β-bounded.

Proof. The assertion of Proposition 7.5 follows from the estimate (7.19). □

7.5 Stability in the Generalised Richardson Model

Unlimited growth of hostility between the countries N_1 and N_2 (no constraints (7.2)) can lead to a real conflict with unpredictable consequences. Let us consider some cases of stability (instability) of the equilibrium state in the model (7.1) with hostility changing over time.

Proposition 7.6. *If in the model (7.1), the hostility functions $g(t, x, y)$ and $h(t, x, y)$ satisfy the conditions $g(t, 0, 0) = 0$ and $h(t, 0, 0) = 0$ for all $t \in \mathbb{R}_\tau$ and the following inequalities hold:*

(1) *$(k(t) + l(t))xy - x(a(t)x - g(t, x, y)) - y(b(t)y - h(t, x, y))$*
   $\leq \theta_1(t)(x^2 + y^2)$
   under $\theta_1(t) > 0$, and

$$\int_{t_r}^\infty \theta_1(s)ds = L_1 < \infty \tag{7.22}$$

   for all $t \in \mathbb{R}_\tau$,
(2) *$(k(t) + l(t))xy - x(a(t)x - g(t, x, y)) - y(b(t)y - h(t, x, y))$*
   $\geq \theta_2(t)(x^2 + y^2)$

under $\theta_2(t) > 0$, and

$$\int_{t_r}^{\infty} \theta_2(s)ds = \infty \qquad (7.23)$$

for all $t \in \mathbb{R}_\tau$,

then the equilibrium state $(x = 0, y = 0)^T$ *is stable (unstable) when the inequalities (7.22) and (7.23) are satisfied.*

Proof. The assertions of Proposition 7.6 are obtained by applying the Lyapunov function $V(x, y) = \frac{1}{2}(x^2 + y^2)$ to analyse the stability (instability) of the equilibrium state $(x = 0, y = 0)^T$ in the system (7.1). Let us show this. The total derivative of the function $V(x, y)$ due to the system of equations (7.1) has the form

$$\frac{d}{dt}V(x, y) = \frac{dx}{dt}x + \frac{dy}{dt}y = x(k(t)y - a(t)x + g(t, x, y))$$
$$+ y(l(t)x - b(t)y + h(t, x, y)). \qquad (7.24)$$

This expression, condition (1) of Proposition 7.6, and Lyapunov's Theorem I on the stability of motion (see [76, pp. 61–62]) imply the first assertion of Proposition 7.6. Similarly, if condition (2) of Proposition 7.6 is satisfied and by Lyapunov's Theorem II on the instability of motion (see [76, pp. 65–66]) the second assertion of Proposition 7.6 is valid. $\qquad \square$

Proposition 7.7. *If in the model (7.1), the hostility functions $g(t, x, y)$ and $h(t, x, y)$ satisfy the conditions $g(t, 0, 0) = 0$ and $h(t, 0, 0) = 0$ for all $t \in \mathbb{R}_\tau$ and, at the moment $t_c \in \mathbb{R}_\tau$, are equivalent in influence on the threat level, i.e., $g(t, x, y) - k(t)y \equiv 0$ and $h(t, x, y) - l(t)x \equiv 0$, then the mathematical model (7.1) degenerates into the following:*

$$\frac{dx}{dt} = -a(t)x, \quad x(t_c) = x_{0c}; \quad \frac{dy}{dt} = -b(t)y, \quad y(t_c) = y_{0c} \qquad (7.25)$$

and the armament *levels of the opposing countries.*

Proof. Using the Lyapunov function of the form $V(x, y) = xy$, it is easy to show that

$$x(t, t_c, x_{0c})y(t, t_c, y_{0c}) = x_{0c}y_{0c}\exp\left(-(a(t) + b(t))(t - t_c)\right) \qquad (7.26)$$

or

$$x(t, t_c, x_{0c}) = x_{0c}\exp\left(-a(t - t_c)\right) \quad \text{and} \quad y(t, t_c, y_{0c}) = y_{0c}\exp\left(-b(t - t_c)\right) \qquad (7.27)$$

for all $t \geq t_c \in \mathbb{R}_\tau$.

This may indicate some kind of emergency situation since in the normal mode of coexistence of opposing countries, such a phenomenon is not observed. $\qquad\square$

Proposition 7.8. *If in the model* (7.1), *the hostility functions are negative,* $-g(t,x,y)$ *and* $-h(t,x,y)$, *and satisfy the conditions* $g(t,0,0) = 0$ *and* $h(t,0,0) = 0$ *for all* $t \in \mathbb{R}_\tau$ *and, at the moment* $t^c \in \mathbb{R}_\tau$, *are equivalent in influence on the level of expenditures by each country on maintaining its armament, i.e.,* $-g(t,x,y) + a(t)x \equiv 0$, *and* $-h(t,x,y) + b(t)y \equiv 0$, *then the mathematical model* (7.1) *becomes*

$$\frac{dx}{dt} = k(t)y, \quad x(t^c) = x_0^c; \quad \frac{dy}{dt} = l(t)x, \quad y(t^c) = y_0^c, \qquad (7.28)$$

and for all $t \geq t^c \in \mathbb{R}_\tau$, *each country determines its level of armament based on the level of armament of the opposite side without taking into account its economic expenditures. In this case, the equilibrium state* $(x = 0, y = 0)^T$ *of the system* (7.1) *is unstable.*

Proof. For the Lyapunov function $V(x,y) = \frac{1}{2}(x^2 + y^2)$ on the solutions of the system (7.28), we have the relation

$$V(x(t), y(t)) = V(x_0^c, y_0^c) \exp(k(t) + l(t))(t - t_c), \qquad (7.29)$$

which implies the assertion of Proposition 7.8. $\qquad\square$

This could correspond to a *global militarisation* of the countries involved in the confrontation.

7.6 Stability in a Model with Autonomous Linear Part

We describe the model of confrontation between the two countries N_1 and N_2 by a system of differential equations with an autonomous linear approximation:

$$\frac{dx}{dt} = ky - ax + g(t,x,y), \quad \frac{dy}{dt} = lx - by + h(t,x,y), \qquad (7.30)$$

where $t \in \mathbb{R}_\tau \subseteq \mathbb{R}_+$ and $x, y \in \mathbb{R}_+$ are measures of the countries' armament levels. Here, the parameters $k > 0$, $l > 0$, $a > 0$, and $b > 0$, and the functions $g(t,x,y)$ and $h(t,x,y)$ have the same meaning as in the model (7.1).

Along with the system of differential equations (7.30), we consider the system of equations

$$\frac{dx}{dt} = ky - ax + \bar{b}_1, \quad \frac{dy}{dt} = lx - by + \bar{b}_2, \qquad (7.31)$$

with hostility level limit values $\bar{b}_1$ and $\bar{b}_2$.

Remark 7.3. If the countries N_1 and N_2 possess nuclear weapons, then the parameters k and l can take arbitrarily large values.

The equilibrium state $(x^e, y^e)^T$ of the system (7.31) is determined from the system of algebraic equations

$$ky - ax + \bar{b}_1 = 0, \quad lx - by + \bar{b}_2 = 0.$$

Lemma 7.2. *If* $\det A \neq 0$, *where* $A = \begin{pmatrix} -a & k \\ l & -b \end{pmatrix}$, *then*

$$(x^e, y^e)^T = -A^{-1}\bar{b}, \tag{7.32}$$

where $\bar{b} = (\bar{b}_1, \bar{b}_2)^T$ *and the superscript* T *denotes transpose.*

Proof. For the proof, see [152, p. 267]. $\qquad\qquad\square$

In the neighbourhood of the equilibrium state (7.32), the system of equations (7.31) is reduced to the form

$$\frac{dw}{dt} = Aw$$

by using the substitution $w = \hat{x} - w^e$, where $\hat{x} = (x, y)^T$ and $w^e = (x^e, y^e)^T$.

Lemma 7.3. *If* $\det A > 0$ *and the condition* (7.2) *is fulfilled, then the time-moving equilibrium state* $(x_e(t), y_e(t))^T$ *in the system* (7.30) *is such that*

$$x_e(t) \leq x^e, \quad y_e(t) \leq y^e \quad \text{for all} \quad t \in \mathbb{R}_\tau.$$

Proof. Since

$$0 = ky_e - ax_e + g(t, x_e, y_e) \leq ky_e - ax_e + |g(t, x_e, y_e)|$$
$$\leq ky_e - ax_e + \bar{b}_1 = ky_e - ax_e - ky^e - ax^e,$$

we get

$$k(y^e - y_e) \leq a(x^e - x_e).$$

Similarly,

$$0 = lx_e - by_e + h(t, x_e, y_e) \leq lx_e - by_e + |h(t, x_e, y_e)|$$
$$\leq lx_e - by_e + \bar{b}_2 = lx_e - by_e - lx^e - by^e.$$

Thus

$$l(x^e - x_e) \leq b(y^e - y_e).$$

Hence,

$$lk(x^e - x_e) \leq bk(y^e - y_e) \leq ab(x^e - x_e),$$

so

$$(x^e - x_e)\det A \geq 0.$$

Similarly,

$$lk(y^e - y_e) \leq al(x^e - x_e) \leq ab(y^e - y_e),$$

so

$$(y^e - y_e)\det A \geq 0.$$

This implies the assertion. □

Since the diagonal elements in the matrix A are negative and the off-diagonal elements are positive, the matrix A is a *Metzler matrix* [126]. We also recall (see [126]) that a matrix A is called a *Hicks matrix* if all odd-order principal minors of A are negative and all even-order principal minors of A are positive.

Next, we need the following fundamental result (see [153]). A Metzler matrix A is stable if and only if it is a Hicks matrix.

Proposition 7.9. *The Metzler matrix A is stable if and only if the conditions*

$$(-1)^r \begin{vmatrix} -a & k \\ l & -b \end{vmatrix} > 0 \quad when \quad r = 1, 2$$

are satisfied.

Proof. For a proof, see [154, p. 403]. □

A Metzler matrix A is a Hicks matrix (see [154] and the bibliography therein) if and only if it is a quasi-dominantly diagonal matrix, i.e., there are constants $d_i > 0$, $i = 1, 2$, such that

$$d_1 a > d_2 k, \quad d_2 b > d_1 l. \tag{7.33}$$

The inequalities (7.33) are sufficient for the stability of the equilibrium state (7.32), limiting the level of hostility of the system (7.31). Since the elements $a_{12} = k$ and $a_{21} = l$ in the matrix A can be arbitrarily large, according to Remark 7.3, to satisfy the inequalities (7.33), the parameters $a = a_{11}$ and $b = a_{22}$, characterising the spending of the countries N_1 and N_2 on armament levels, should be adequate.

Let us return to the system (7.31) and assume that $a_{11} > 0$. This condition implies a profit from the sale of arms by the country N_1. In this case, the matrix A ceases to be Metzler, and the Hicks condition is not fulfilled. As a result, we obtain instability of the equilibrium state (7.32) in the system (7.31).

If the condition $a_{11}a_{22} - a_{12}a_{21} < 0$ is fulfilled in the system (7.31), which corresponds to the situation "fatigue is less aggressive", then the equilibrium state (7.32) in the system (7.31) is unstable.

It is clear that the instability of the equilibrium state in the mathematical model (7.31) of the confrontation between two countries may indicate an approach to conflict.

Remark 7.4. Unlimited growth of hostility between the countries N_1 and N_2 (lack of restrictions (7.2)) can lead to a real conflict between nuclear countries with unpredictable consequences.

Proposition 7.10. *If in the model* (7.30), *the hostility functions* $g(t, x, y)$ *and* $h(t, x, y)$ *satisfy the conditions* $g(t, 0, 0) = h(t, 0, 0) = 0$ *for all* $t \in R_\tau$ *and*

$$(k + l)xy - x(ax - g(t, x, y)) - y(by - h(t, x, y)) \leq \theta_3(t)(x^2 + y^2)$$

under $\theta_3(t) > 0$, *and*

$$\int_0^\infty \theta_3(s)ds = L_2 < \infty,$$

then the equilibrium state $(x = 0, \ y = 0)^T$ *is stable.*

Proof. Consider the Lyapunov function $2V(x, y) = x^2 + y^2$ for the system of equations (7.30). The function $V(x, y)$ is definitely positive, and under the conditions of Proposition 7.10, its derivative is non-increasing on the solutions of the system (7.30). Therefore, based on Lyapunov's first theorem (see [76], p. 62), we obtain that the equilibrium state $x = y = 0$ of system (7.30) is stable. $\qquad\qquad\square$

Proposition 7.11. *If in the model* (7.30), *the hostility functions* $g(t, x, y)$ *and* $h(t, x, y)$ *satisfy the conditions* $g(t, 0, 0) = h(t, 0, 0) = 0$ *for all* $t \in R_\tau$ *and, at time* $t_c \in R_\tau$, *are equivalent in influence on the threat level, i.e.,*

$$g(t, x, y) - ky \equiv 0 \quad and \quad h(t, x, y) - lx \equiv 0,$$

then the system of equations (7.30) *degenerates into*

$$\frac{dx}{dt} = -ax, \quad x(t_c) = x_{0c}; \quad \frac{dy}{dt} = -by, \quad y(t_c) = y_{0c},$$

and the levels of armament of the opposing countries decrease exponentially:

$$x(t, t_c, x_{0c}) = x_{0c} \exp\left(-a(t - t_c)\right) \quad and$$

$$y(t, t_c, y_{0c}) = y_{0c} \exp\left(-b(t - t_c)\right)$$

for all $t \geq t_c \in \mathbb{R}_\tau$.

This may indicate some kind of emergency since in the normal mode of coexistence of opposing countries, such a phenomenon is not observed.

Proposition 7.12. *If the negative hostility functions $-g(t, x, y)$ and $-h(t, x, y)$ in the model (7.30) satisfy the conditions $g(t, 0, 0) = h(t, 0, 0) = 0$ for all $t \in \mathbb{R}_\tau$ and, at time $t^c \in \mathbb{R}_\tau$, maintain their armament, i.e.,*

$$-g(t, x, y) + ax \equiv 0 \quad and \quad -h(t, x, y) + by \equiv 0,$$

then the system of equations (7.30) degenerates into

$$\frac{dx}{dt} = ky, \quad x(t^c) = x_0^c; \quad \frac{dy}{dt} = lx, \quad y(t^c) = y_0^c,$$

and for all $t \geq t^c \in \mathbb{R}_\tau$, each of the countries determines its level of armament based on the level of armament of the opposite side, not taking into account economic costs.

Proof. This assertion follows from the explicit form of the levels of armament of the countries involved in the confrontation that are determined by the expressions

$$x(t, t^c, x_0^c, y_0^c) = \left(\frac{k}{l}\right)^{\frac{1}{2}} \left(C_1 \exp\left(\lambda_1(t - t^c)\right) - C_2 \exp\left(-\lambda_2(t - t^c)\right)\right),$$

$$y(t, t^c, x_0^c, y_0^c) = C_1 \exp\left(\lambda_1(t - t^c)\right) + C_2 \exp\left(-\lambda_2(t - t^c)\right)$$

for all $t \geq t_c \in \mathbb{R}_\tau$, where

$$\lambda_{1,2} = \pm(kl)^{\frac{1}{2}}, \quad C_1 = \frac{y_0^c + \left(\frac{l}{k}\right)^{\frac{1}{2}} x_0^c}{2}, \quad C_2 = \frac{y_0^c - \left(\frac{l}{k}\right)^{\frac{1}{2}} x_0^c}{2}.$$

This case may correspond to a global militarisation of the countries involved in the confrontation. □

Proposition 7.13. *If in the model (7.30), the hostility functions $g(t, x, y)$ and $h(t, x, y)$, at the moment $\bar{t} \in \mathbb{R}_\tau$, satisfy the conditions (7.2), then the system of equations (7.30) takes the form (7.31), and for the armament*

*levels $x(\bar{t}) = \bar{x}_0$ and $y(\bar{t}) = \bar{y}_0$, the countries' armament levels for all $t \geq \bar{t}$
are represented (cf. [171]) as*

$$x(t, \bar{t}, \bar{x}_0, \bar{y}_0) = x_1 + \frac{kl}{2\mu(\mu + \omega)} \bar{C}_1 \exp(\lambda_1(t - \bar{t}))$$

$$+ \frac{\mu + \omega}{2\mu} \bar{C}_2 \exp(\lambda_2(t - \bar{t})), \qquad (7.34)$$

$$y(t, \bar{t}, \bar{x}_0, \bar{y}_0) = y_1 + \frac{l}{2\mu} \bar{C}_1 \exp(\lambda_1(t - \bar{t}))$$

$$+ \frac{1}{2\mu} \bar{C}_2 \exp(-\lambda_2(t - \bar{t})) \qquad (7.35)$$

for all $t \geq \bar{t} \in \mathbb{R}_\tau$. Here, $\omega = \frac{1}{2}(a - b)$, $\mu = (\omega^2 + kl)^{\frac{1}{2}}$,

$$x_1 = \frac{b\bar{b}_1 + \bar{b}_2 k}{ab - kl}, \quad y_1 = \frac{a\bar{b}_2 + \bar{b}_1 l}{ab - kl}, \quad \lambda_{1,2} = -\frac{1}{2}(a + b) \pm ((a - b)^2 + 4kl)^{\frac{1}{2}},$$

$$\bar{C}_1 = \left(\bar{x}_0 + \frac{\bar{b}_1}{\lambda_1}\right) + \frac{\mu + \omega}{l}\left(\bar{y}_0 + \frac{\bar{b}_2}{\lambda_1}\right),$$

$$\bar{C}_2 = \left(\bar{x}_0 + \frac{\bar{b}_1}{\lambda_1}\right) - \frac{k}{\mu + \omega}\left(\bar{y}_0 + \frac{\bar{b}_2}{\lambda_2}\right).$$

For some numerical values of the parameters of the model (7.31), the solutions (7.34) and (7.35) allow us to observe the dynamics of armament levels in supercritical time for $t \geq \bar{t}$.

7.7 Analysis of the Model of Confrontation between n Countries Involved in Alliances

Let us assume that the armament of n countries changes over time under the influence of scientific and technological progress and growing hostility between the countries.

Let us assume that the system of differential equations in the form

$$\dot{x}(t) = \frac{dx}{dt} = A(t, x)x + b(t, x), \quad x(t_0) = x_0 \qquad (7.36)$$

simulates the described situation. Here, $x \in \mathbb{R}_+^n$ is the armament vector of n countries, $A(t, x)$ is an $n \times n$ matrix that characterises the armament of n countries involved in alliances, and $b(t, x)$ is a vector function, nonlinear in x, which characterises public opinion about the arms race and hostility between n countries.

We assume that the matrix $A(t, x)$ and the vector function $b(t, x)$ are defined, bounded, and continuous on the product $\mathbb{R}_\tau \times \mathbb{R}_+^n$ so that, under the initial conditions $(t_0, x_0) \in \mathbb{R}_\tau \times \mathbb{R}_+^n$, the solution to the initial problem (7.36) exists for all $t \in \mathbb{R}_\tau$, $\mathbb{R}_\tau \subseteq \mathbb{R}_+$.

Let, for some value $t^* \in \mathbb{R}_\tau$, the armament of alliances and hostility between countries be characterised by the matrix $\bar{A}$ and the vector $\bar{b}$ such that the inequality $A(t, x) \leq \bar{A}$ is satisfied element by element for the matrix $A(t, x)$ and the inequality $b(t, x) \leq \bar{b}$ is fulfilled component by component for the vector $\bar{b}$ for all $t \geq t^* \in \mathbb{R}_\tau$ and $x \in \mathbb{R}_+^n$.

Along with the system of equations (7.36), we consider the system of differential equations simulating the boundary situation of confrontation in the form

$$\frac{dx}{dt} = \bar{A}x + \bar{b}, \quad x(t^*) = x_0^*. \tag{7.37}$$

The equilibrium state of the system (7.37) is determined by the formula

$$(x^e)^T = -\bar{A}^{-1}\bar{b}$$

provided that $\det \bar{A} \neq 0$.

The moving equilibrium state $(x_e(t))^T$ of the system (7.36) is determined from the system of algebraic equations

$$A(t, x)x + b(t, x) = 0 \quad \text{for all} \quad t \in \mathbb{R}_\tau.$$

Lemma 7.4. *If $\det \bar{A} > 0$ and condition $b(t, x) \leq \bar{b}$ is fulfilled, then the moving equilibrium $x_e(t) \in \mathbb{R}_+^n$ of the system (7.36) satisfies the condition*

$$x_e(t) \leq x^e \quad \text{for all} \quad t \geq t^* \in \mathbb{R}_\tau.$$

Proof. The proof is similar to that of Lemma 7.3, and for this reason, it is omitted here. $\qquad\square$

Next, we consider the behaviour of solutions to the system (7.36) in the neighbourhood of the moving equilibrium state $x_e(t) \in \mathbb{R}_+^n$.

Further, for the mathematical model (7.36) of confrontation between n countries, we consider two problems.

Problem A. Suppose that in the model (7.36), the hostility vector function $b(t, x)$ does not lower the level of the protection factor. It is necessary to establish conditions for the decrease in the norm of the armament vector of n countries involved in alliances.

Suppose that the countries involved in the confrontation form alliances, which are described by the matrix of the fundamental interactions E (see [154, pp. 8, 66, 273]). The elements $e_{ij}(t)$, $i, j = 1, 2 \ldots, n$, of the matrix E are defined by

$$e_{ij}(t) = \begin{cases} 1 & \text{if } x_j \text{ interacts with } x_i, \\ 0 & \text{if } x_j \text{ does not interact with } x_i. \end{cases}$$

Suppose that the elements $a_{ij}(t, x)$ of the matrix $A(t, x)$ are of the (see [153]) form

$$a_{ij}(t, x) = \begin{cases} -(\alpha_i(t, x) + e_{ii}(t)\alpha_{ii}(t, x)) & \text{for } i = j, \\ e_{ij}(t)\alpha_{ij}(t, x) & \text{for } i \neq j. \end{cases} \tag{7.38}$$

Here, $\alpha_i(t, x) \geq \overline{\alpha}_i$, $|\alpha_{ij}(t, x)x_j| \leq \overline{\alpha}_{ij}|x_j|$, $i, j = 1, 2, \ldots, n$, $\overline{\alpha}_i < \overline{\alpha}_{ii}$, and $x = (x_1, x_2, \ldots, x_n)^T$ are the armament vectors, and $e_{ij}(t) \in E$, $\overline{\alpha}_i > 0$, and $\overline{\alpha}_{ij} \geq 0$ are some constants.

If $0 \leq e_{ij}(t) \leq 1$ for all $t \in \mathbb{R}_\tau$, then the fundamental interaction matrix E is denoted by $\overline{E} = (\overline{e}_{ij})$, and it is assumed that the zeros of the matrix E remain the zeros of the time-invariant matrix $\overline{E}$.

The elements of $\overline{E}$ take on a binary value, say 1, if country j with the armament level x_j interacts with the country i with the armament level x_i, and 0 if country j does not interact with country i.

Let us establish the conditions for a decrease in the norm of the armament vector in the neighbourhood of the equilibrium state $x_e(t) \in \mathbb{R}_+^n$. To this end, we use the method of Lyapunov functions. Let the function $V(x)$ be of the form

$$V(x) = \sum_{i=1}^{n} x_i^2 = x^T x. \tag{7.39}$$

For the right Dini derivative of the function (7.39) to the system (7.36), we obtain

$$D^+ V(x(t)) \equiv \dot{x}^T x + x^T \dot{x}$$
$$= \left(x^T A^T(t, x) + b^T(t, x)\right) x + x^T \left(A(t, x)x + b(t, x)\right)$$
$$= x^T \left(A^T(t, x) + A(t, x)\right) x + 2b^T(t, x)x = x^T G(t, x)x + 2b^T(t, x)x,$$

where $G(t, x) = A^T(t, x) + A(t, x)$.

Suppose that for the matrix $G(t, x)$, there exists a negative-definite constant matrix $\bar{G}$ such that the inequality $G(t, x) \leq \bar{G}$ holds for all $t \in \mathbb{R}_\tau$. The quadratic form $x^T \bar{G} x$ satisfies the relation

$$\lambda_m(\bar{G}) x^T x \leq x^T \bar{G} x \leq \lambda_M(\bar{G}) x^T x, \quad \lambda_M(\bar{G}) < 0.$$

To estimate the *maximum level of hostility*, we introduce (cf. [81, p. 28]) the function

$$R(t) = \max_{x \in \mathbb{R}_+^n} \frac{2b^T(t, x)x}{x^T x}.$$

Taking into account the accepted notation for $D^+V(x(t))$, we obtain the estimate

$$D^+V(x(t, t_0, x_0)) \leq -\gamma(t)V(x(t, t_0, x_0)) \quad \text{for all} \quad t \in \mathbb{R}_\tau,$$

where $\gamma(t) = -\lambda_M(\bar{G}) + R(t)$. If for all $t \in \mathbb{R}_\tau$, the inequalities $-\lambda_M(\bar{G}) + R(t) < 0$ are fulfilled, then for the function $V(x(t))$, we obtain the inequality

$$V(x(t, t_0, x_0)) \leq V(x_0) \exp\left(-\int_{t_0}^t \gamma(s)ds\right)$$

for all $t \in \mathbb{R}_\tau$ and $(t_0, x_0) \in \mathbb{R}_+ \times \mathbb{R}^n$. Given that $\|x\| = \left(\sum_{i=1}^n x_i^2\right)^{1/2}$, we obtain an estimate for the norm of the vector $x(t)$ as

$$\|x(t, t_0, x_0)\| \leq \|x_0\| \exp\left(-\frac{1}{2}\int_{t_0}^t \gamma(s)ds\right) \tag{7.40}$$

for all $t \in \mathbb{R}_\tau$ and $(t_0, x_0) \in \mathbb{R}_+ \times \mathbb{R}^n$.

Now, we can formulate the following proposition.

Proposition 7.14. *If for all $t \in \mathbb{R}_\tau$, all conditions are satisfied for the right-hand side of the system (7.36) with the matrix (7.38) and the function $V(x(t))$ and, moreover, for all $t \in \mathbb{R}_\tau$, the condition*

$$-\lambda_M(\bar{G})(t - t_0) + \int_{t_0}^t R(s)ds < 0$$

is satisfied, then the measure of hostility $R(t)$ between the countries involved in the confrontation does not affect the decrease in the norm of the armament vector. In this case, the estimate of the decrease in the norm of the armament vector is expressed by the formula (7.40).

Remark 7.5. If there exists $t_1 \in \mathbb{R}_\tau$ such that $-\lambda_M(\bar{G})(t - t_1) + \int_{t_1}^t R(s)ds \equiv 0$ holds for all values $t > t_1 \in \mathbb{R}_\tau$, then for all $t > t_1$, the decrease in the norm of the armament vector stops.

Problem B. Let us assume that in the model (7.36), the hostility vector function describes the weakening of the protection factor, i.e., in the system (7.36), the vector function $b(t, x)$ is negative for all $t \in \mathbb{R}_\tau$. It is necessary to establish the conditions for the growth of the norm of the armament vector of n countries involved in alliances.

To obtain the growth conditions for the armament vector norm, consider the model (7.36). Suppose that the elements of the matrix $A(t, x)$ are of the form (see [153])

$$c_{ij}(t, x) = \begin{cases} \varphi_i(t, x) - e_{ij}(t)\varphi_{ii}(t, x) & \text{for } i = j, \\ e_{ij}(t)\varphi_{ij}(t, x) & \text{for } i \neq j, \end{cases}$$

where the functions $\varphi_i(t, x)$, $\varphi_{ii}(t, x)$, and $\varphi_{ij}(t, x)$ satisfy the same conditions as the elements in the matrix (7.38). In this case, the elements of the matrix $\overline{A}$ are defined as

$$\overline{c}_{ij} = \begin{cases} \overline{\alpha}_i - \overline{e}_{ii}\overline{\alpha}_{ii} & \text{for } i = j, \\ \overline{e}_{ij}\overline{\alpha}_{ij} & \text{for } i \neq j. \end{cases}$$

For the Dini derivative of the function (7.39) on the system's solutions

$$\dot{x}(t) = \frac{dx}{dt} = A(t, x)x - b(t, x), \quad x(t_0) = x_0, \tag{7.41}$$

it is easy to obtain an estimate:

$$\begin{aligned} D^+V(x(t)) &= \left(x^T A^T(t, x) - b^T(t, x)\right)x + x^T\left(A(t, x)x - b(t, x)\right) \\ &= x^T\left(A^T(t, x) + A(t, x)\right)x - 2b^T(t, x)x \\ &= x^T C(t, x)x - 2b^T(t, x)x, \end{aligned}$$

where $C(t, x) = A^T(t, x) + A(t, x)$ for all $(t, x) \in \mathbb{R}_+ \times \mathbb{R}^n$.

Suppose that for the matrix $C(t, x)$, there exists a positive-definite constant matrix $\overline{C}$ such that the inequality $C(t, x) \geq \overline{C}$ holds for all $t \in \mathbb{R}_\tau$. The quadratic form $x^T\overline{C}x$ satisfies the inequality

$$\lambda_m(\overline{C})x^T x \leq x^T\overline{C}x \leq \lambda_M(\overline{C})x^T x, \quad \lambda_M(\overline{C}) > 0.$$

The *minimum level of hostility* is characterised by the function

$$Q(t) = \min_{x \in \mathbb{R}_+^n} \frac{2b^T(t, x)x}{x^T x}.$$

Taking into account the accepted notation for the Dini derivative $D^+V(x(t))$, we obtain the estimate

$$D^+V(x(t, t_0, x_0)) \geq \pi(t)V(x(t, t_0, x_0)) \quad \text{for all} \quad t \in \mathbb{R}_\tau,$$

where $\pi(t) = \lambda_m(\bar{C}) - Q(t)$. If for all $t \in \mathbb{R}_\tau$, the inequalities $\lambda_m(\bar{C}) - Q(t) > 0$ are fulfilled, then for the function $V(x(t))$, we obtain the inequality

$$V(x(t, t_0, x_0)) \geq V(x_0) \exp\left(\int_{t_0}^{t} \pi(s)ds\right)$$

for all $t \in \mathbb{R}_\tau$ and $(t_0, x_0) \in \mathbb{R}_+ \times \mathbb{R}^n$. Given that $\|x\| = \left(\sum_{i=1}^{n} x_i^2\right)^{1/2}$, we obtain an estimate for the norm of the vector $x(t)$ as

$$\|x(t, t_0, x_0)\| \geq \|x_0\| \exp\left(\frac{1}{2}\int_{t_0}^{t} \pi(s)ds\right) \tag{7.42}$$

for all $t \in \mathbb{R}_\tau$ and $(t_0, x_0) \in \mathbb{R}_+ \times \mathbb{R}^n$.

Proposition 7.15. *If the condition $\lambda_m(\bar{C})(t - t_0) - \int_{t_0}^{t} Q(s)ds > 0$ is satisfied for all $t \geq t_0$, then the measure of hostility $Q(t)$ between the countries involved in the confrontation does not affect the growth of the hostility vector. In this case, the estimate of the increases in the norm of the armament vector is expressed by the formula (7.42).*

Remark 7.6. The inequality (7.42) indicates that, even with a negative opinion about the arms race, the level of armament of the countries involved in the confrontation will increase. From the inequality $\lambda_m(\bar{C})(t - t_0) - \int_{t_0}^{t} Q(s)ds > 0$, it follows that if the parameters of the protection factor are weakened by hostility, then the role of alliances should increase.

Remark 7.7. The estimate (7.42) implies that, if for all values $t > t_1 \in \mathbb{R}_\tau$, the hostility measure is such that $\lambda_m(\bar{C})(t - t_1) - \int_{t_1}^{t} Q(s)ds \equiv 0$, then there is no increase in the norm of the armament vector for all $t \geq t_1$.

7.8 Nonlinear Confrontation Model

To describe a general model of confrontation between n countries, we use the system of differential equations

$$\frac{dx}{dt} = H(t, x) + B(t, x), \tag{7.43}$$

where $x \in \mathbb{R}_+^n$ is the armament vector, $H : \mathbb{R}_+ \times \mathbb{R}_+^n \to \mathbb{R}^n$ is a vector function describing the level of armament of all countries involved in the confrontation, and $B(t, x)$ is a vector function of hostility between all countries, depending on their armament, linear in x. As in the Richardson model, we assume that in the model (7.43), the growth of the armament of one country depends on the growth of the armament of the other countries involved in the confrontation. This assumption is formalised as

$$H_i(t, a) < H_i(t, \tilde{a}) \text{ for all } (t, a), (t, \tilde{a}) \in \mathbb{R}_+ \times \mathbb{R}_+^n,$$

$$a_i = \tilde{a}_i, \quad a_j < \tilde{a}_j, \quad i \in M, \quad j \in N \backslash M, \tag{7.44}$$

where N is the set of indices $\{1, 2, \ldots, n\}$ and M is a nonempty subset of N. In addition, we assume that for any $\lambda > 0$, the vector function $H(t, x)$ satisfies (see [30, 153]) the condition

$$H(t, \lambda x) = H(t, x). \tag{7.45}$$

Let the system of equations (7.43) have a positive equilibrium state x_e, i.e.,

$$H(t, x_e) + B(t, x_e) = 0 \tag{7.46}$$

such that $x_e \in \Delta$, where $\Delta = \{x \in \mathbb{R}_+^n : x > 0\}$ is an open cone in $\mathbb{R}_+^n$.

Lemma 7.5. *If the conditions* (7.44)–(7.46) *are satisfied for the system of equations* (7.43), *then there is a single equilibrium ray* $L = \{x_e \in \Delta : x_e = \lambda e\}$, *where* $e = (1, 1, \ldots, 1) \in \mathbb{R}_+^n$ *and* λ *is a positive number.*

Proof. For a proof, see [152, Lemma 1]. □

We introduce the formula $d[x, L] = \inf\limits_{x^e \in L} \|x - x^e\|_N$, where $\|x\|_N = \sup\limits_{i \in N} \|x_i\|$ is the distance between the armament vector and the equilibrium ray, and we assume that the vector functions $H(t, x)$ and $B(t, x)$ in the system of equations (7.43) satisfy the following conditions. There are nonnegative functions $f_1(t)$ and $f_2(t)$ such that:

(a) $d[H(t, x), L] \le f_1(t) d[x, L]$ for all $t \in \mathbb{R}_\tau$, and $x \in \mathbb{R}_+^n$;
(b) $d[B(t, x), L] \le f_2(t) d^\alpha[x, L]$ for all $t \in \mathbb{R}_\tau$, $\alpha > 1$, and $x \in \mathbb{R}_+^n$.

Let us establish an estimate for the deviation of the armament vector $x(t)$ from the equilibrium ray L in the system (7.43) on the basis of a nonlinear integral inequality. Let us rewrite the system of equations (7.43) as

$$x(t, t_0, x_0) = x_0 + \int_{t_0}^t (H(s, x(s)) + B(s, x(s))) ds.$$

Taking estimates (a) and (b) into account, we obtain

$$d[x(t, t_0, x_0), L] = d[x_0, L] + \int_{t_0}^t d[(H(s, x(s)) + B(s, x(s))), L] ds$$

$$\le d[x_0, L] + \int_{t_0}^t (f_1(s) d[x(s, t_0, x_0), L] + f_2(s) d^\alpha[x(s, t_0, x_0), L]) ds.$$

$$\tag{7.47}$$

Since $x_0 > 0$, we get $d[x_0, \mathrm{L}] \geq 0$. Let us assume that the condition

$$\Phi(t_0, t) = (\alpha - 1)d^{\alpha-1}[x_0, \mathrm{L}] \int_{t_0}^{t} f_2(s) \exp\left((\alpha - 1)\int_{t_0}^{s} f_1(\tau)d\tau\right) ds < 1 \tag{7.48}$$

holds for all $t \in \mathbb{R}_\tau$.

The deviation of the armament vector from the equilibrium ray is estimated using the inequality

$$d[x(t, t_0, x_0), \mathrm{L}] \leq d[x_0, \mathrm{L}] \exp\left(\int_{t_0}^{t} f_1(s)ds\right)(1 - \Phi(t_0, t))^{-\frac{1}{\alpha-1}} \tag{7.49}$$

for all t, for which the condition (7.48) is satisfied. The bound (7.49) is obtained by applying the nonlinear integral inequality (see [83, Theorem 2, pp. 185–186] and the bibliography therein) to the inequality (7.47).

Proposition 7.16. *If for any $\varepsilon > 0$, there exists $\delta = \delta(t_0, \varepsilon) > 0$ such that:*

(a) $d[x_0, \mathrm{L}] < \delta$;

(b) $\overline{\Phi}(t_0, t) = (\alpha - 1)\delta^{\alpha-1} \int_{t_0}^{t} f_2(s) \exp\left((\alpha - 1)\int_{t_0}^{s} f_1(\tau)d\tau\right) ds < 1$ *for all* $t \in \mathbb{R}_\tau$;

(c) $\exp\left(\int_{t_0}^{t} f_2(s)ds\right)\left(1 - \overline{\Phi}(t_0, t)\right)^{-\frac{1}{\alpha-1}} < \frac{\varepsilon}{\delta}$ *for all* $t \in \mathbb{R}_\tau$,

then

$$d[x(t, t_0, x_0), \mathrm{L}] < \varepsilon$$

for all t, for which conditions (b) and (c) are satisfied.

Proof.   The proof follows from the estimate (7.49).   □

Thus, conditions (a)–(c) are sufficient for the stability of the deviation of the armament vector of n countries in the model (7.43) from the equilibrium ray in the form (7.49).

7.9   Comments and Bibliography

The current state of scientific and technical development, particularly in the field of nuclear physics, is characterised by the availability of nuclear weapons to many countries of the world (in addition to those countries that already possess them). This increases tension among states and changes the overall security structure in the world as a whole. Unfortunately, the Russell–Einstein Manifesto of 1955 (available on the internet) seems to have been forgotten.

Consequently, any war (as an anachronism of past centuries) involving countries possessing nuclear weapons against countries that do not have such weapons has no rational explanation. In these conditions, maintaining the balance between the opposing nuclear countries remains one of the most important problems.

Note that by maintaining the stability of the equilibrium state in the process of nuclear confrontation, we mean a decrease in the norm of the armament vector of the countries involved in confrontation and the non-use of nuclear weapons both between countries that possess such weapons and against those countries that do not have such weapons.

The collapse of the USSR and the emergence of new states on its territory upset the balance of countries opposing on the European continent. This led to a deterioration in the general security in both Europe and the world after World War II (see Hauser [49], etc.).

In this context, the relevance of research in the field of stability of the balance of opposing countries and their alliances, including the analysis of the strategy of countries possessing nuclear weapons, increases significantly (see Freedman and Michaels [42]).

In the work by Bohner and Martynyuk [21], mathematical models of confrontation between two and n countries, including countries possessing nuclear weapons, are proposed and analysed. The proposed models are based on a generalisation of the well-known Richardson mathematical model of the arms race. As a result of the analysis, the conditions for the stability of the equilibrium state of the opposing countries are established, and the influence of hostility on the decrease (increase) in the norm of the armament vector of n countries participating in alliances is shown.

This chapter considered a non-autonomous quasi-linear model of confrontation between two countries and the hostility factor between them is filled with expanded content. The analysis of the proposed mathematical model of confrontation is carried out using the Lyapunov function method (see Lyapunov [76]) and a nonlinear integral inequality (see Martynyuk [83], N'Doye [124]).

The paper by Smith [161] provides an overview of the research conducted in recent years on the development of the Richardson arms race model. Our book proposes to consider new models of confrontation between two or more countries involved in the confrontation process. These mathematical models are based on some ideas from the papers by Martynyuk [78], Richardson [145], and Siljak [152, 153]. For other recent works on the Richardson model, we refer to Metz and Viorel [113], Gleditsch [44], and Pandey and

Raturi [133]. For related analysis of the stability of two-dimensional systems in various application areas, we refer to Bohner and Martynyuk [22], Martynyuk [79], Pal and Mahapatra [130], Pal, Mahapatra, and Samanta [131, 132], and Santra and Mahapatra [151].

The chapter is based on the results reported in the paper by Bohner and Martynyuk [21], with some development.

As for the prospect of further research in this direction, it will be of interest to consider the mathematical models (7.30), (7.36), and (7.43) under any of the following assumptions:

- the dependence of all parameters of the linear approximation on the level of armaments,
- the fluctuations of all parameters of the linear approximation,
- the presence of a delay in the functions describing hostility,
- the presence of impulse disturbances in the functions describing hostility,
- the presence of control in the functions describing hostility,
- the assessment of the potential impact of artificial intelligence on the formation of hostility between countries.

Moreover, another interesting problem is the consideration of the models (7.30), (7.36), and (7.43) in the discrete case or the so-called time scale setting; see, e.g., Bohner and Martynyuk [22] and Martynyuk [79].

Appendix A

Elements of interval analysis. When studying the motion of systems with interval uncertainty of initial conditions, incomplete (partial) knowledge of solutions of the corresponding system of perturbed motion equations is assumed, i.e., in this situation, we can only indicate the boundaries of the possible values of the solution norm. Accordingly, interval analysis is a branch of mathematics that studies problems with interval uncertainties and methods for solving them. Let us present some information from interval analysis (see [5, 156] and the bibliography therein).

An interval $\boldsymbol{a}$ is a set of the form

$$\boldsymbol{a} := [\underline{a}, \overline{a}] = \{x \in \mathbb{R} | \underline{a} \le x \le \overline{a}\}. \tag{A.1}$$

The real numbers a are identified with zero-width intervals $[a, a]$, also called degenerate intervals. If $\boldsymbol{a}$ is an interval, then $(-\boldsymbol{a})$ is the interval $(-1) \cdot \boldsymbol{a}$, so $-\boldsymbol{a} = [-\overline{a}, -\underline{a}]$. On the set of intervals of form (1), interval arithmetic operations are introduced:

(1) addition:

$$\boldsymbol{a} + \boldsymbol{b} = [\underline{a} + \underline{b}, \overline{a} + \overline{b}]. \tag{A.2}$$

(2) subtraction:

$$\boldsymbol{a} - \boldsymbol{b} = [\underline{a} - \overline{b}, \overline{a} - \underline{b}]. \tag{A.3}$$

(3) multiplication:

$$\boldsymbol{a} \cdot \boldsymbol{b} = [\min\{\underline{a}\underline{b}, \underline{a}\overline{b}, \overline{a}\underline{b}, \overline{a}\overline{b}\}, \max\{\underline{a}\underline{b}, \underline{a}\overline{b}, \overline{a}\underline{b}, \overline{a}\overline{b}\}]. \tag{A.4}$$

(4) division:

$$\boldsymbol{a}/\boldsymbol{b} = \boldsymbol{a} \cdot 1/\overline{b}, 1/\underline{b}, \quad \text{for} \quad 0 \notin \boldsymbol{b}. \tag{A.5}$$

Definition A.1. The algebraic system $\langle \mathbb{IR}, +, -, \cdot, / \rangle$ formed by the set of all real intervals (1) with the binary operations of addition, subtraction, multiplication, and division, which are defined by formulas (2)–(5), is called the classical interval arithmetic.

Definition A.2. The absolute value (modulus) of the interval a is the value $|a| = \max\{|\underline{a}|, |\overline{a}|\}$.

Definition A.3. An ordered tuple of intervals arranged vertically (column vector) or horizontally (row vector) is called an interval vector.

Thus, if a_1 and $a_2, \ldots, a_n$ are some intervals, then $a = (a_1, a_2, \ldots, a_n)^T$ is an interval column vector, and $a = (a_1, a_2, \ldots, a_n)$ is an interval row vector.

The set of interval vectors whose components belong to $\mathbb{IR}$ is denoted by $\mathbb{IR}^n$. The geometric image of an interval vector is a rectangular parallelepiped in the space $\mathbb{R}^n$ with sides parallel to the coordinate axes – a bar.

Definition A.4. The norm of an interval vector a is a real value, denoted by $\|a\|$, that satisfies the following axioms:

(1) $\|a\| \geq 0$, and $\|a\| = 0 \Leftrightarrow a = 0$ – non-negativity;
(2) $\|\alpha a\| = |\alpha| \cdot \|a\|$, for $\alpha \in \mathbb{R}$ – absolute continuity;
(3) $\|a + b\| \leq \|a\| + \|b\|$ – triangle inequality;
(4) $a \subseteq b \Rightarrow \|a\| \leq \|b\|$ – inclusion monotonicity.

The following expressions are examples of interval vector norms:

$$\|a\|_1 := |a_1| + |a_2| + \cdots + |a_n|, \tag{A.6}$$

$$\|a\|_\infty := \max_{1 \leq i \leq n} |a_i|, \tag{A.7}$$

$$\|a\|_2 := \sqrt{|a_1|^2 + |a_2|^2 + \cdots + |a_n|^2}. \tag{A.8}$$

Appendix B

An estimate for the exponent. In many cases of qualitative analysis of quasi-linear systems, it is important to have an effective estimate for the constant $\widetilde{N}$ in an inequality of the type

$$\|e^{Ht}\| \leq \widetilde{N}e^{\sigma t}, \quad t \in [0, \infty),$$

where $\widetilde{N} > 0$ and $\sigma < 0$. Taking into account the results of [28, 165], we indicate a possible method for such an estimate. Let $\rho_j = \alpha_j + i\beta_j$ be the eigenvalues of the matrix H and $\alpha_0 = \max\{\alpha_1, \ldots, \alpha_n\}$. Then, we have the estimate

$$\|e^{Ht}\| \leq \varphi(t)e^{\alpha_0 t}, \tag{B.1}$$

where $\varphi(t) = \sum_{k=0}^{n-1} \frac{(2\|H\|)^k}{k!}t^k$ is a polynomial in t of degree $n-1$. Inequality (B.1) can be rewritten in the form

$$\|e^{Ht}\| \leq \xi(t)e^{(\alpha_0+\alpha)t}, \quad \xi(t) = \varphi(t)e^{-\alpha t}.$$

Let $\alpha_0 < 0$ and $\alpha > 0$ be chosen so that $\alpha_0 + \alpha < 0$. Then, the function $\xi(t) \to 0$ as $t \to \infty$; therefore, $\xi(t)$ has the largest value on the semiaxis $[0, \infty)$, which can be calculated using the usual rules for finding the extremum of a differentiable function. The following lemma shows how to do this.

Lemma B.1. [10] *Under the assumptions made above, the following estimate is valid:*

$$\|e^{Ht}\| \leq F(s^*)e^{(\alpha_0+\alpha)t}, \quad t \in [0, \infty), \tag{B.2}$$

where s^* is the unique solution of the equation

$$\psi_{n-2}(s) = \theta\psi_{n-1}(s) \tag{B.3}$$

on the interval $[0, \infty)$, $\theta = \frac{\alpha}{2\|H\|}$, $\psi_{n-1}(s) = \sum_{k=0}^{n-1} \frac{s^k}{k!}$, $F(s) = \psi_{n-1}(s)e^{-\theta s}$.

Proof.   Before proceeding to the proof of the lemma, we examine the correctness of the definition of θ. Since the largest of the real parts of the eigenvalues of the matrix H, the value of α_0, is negative, we have $\|H\| \neq 0$, so θ is well-defined.   □

In order to obtain estimate (B.2), we find the largest value of the function $\xi(t)$ on the interval $[0, \infty)$. To simplify the calculations, in the function $\xi(t)$, we change the variable t according to the formula $t = \frac{\theta s}{\alpha}$; as a result, we get

$$\xi(t) = \varphi(t)e^{-\alpha t} = \varphi\left(\frac{\theta s}{\alpha}\right)e^{-\theta s} = \psi_{n-1}(s)e^{-\theta s} = F(s).$$

Then, $\max_{[0,\infty)} \xi(t) = \max_{[0,\infty)} F(s)$. To find the largest value of the function $F(s)$ on the semiaxis $[0, \infty)$, we calculate $\frac{dF}{ds}$ as

$$\frac{dF}{ds} = \frac{d\psi_{n-1}(s)}{ds}e^{-\theta s} - \theta\psi_{n-1}(s)e^{-\theta s} = \left(\frac{d\psi_{n-1}(s)}{ds} - \theta\psi_{n-1}(s)\right)e^{-\theta s}.$$

Therefore, the stationary points of the function $F(s)$ are solutions of the equation

$$\frac{d\psi_{n-1}(s)}{ds} = \theta\psi_{n-1}(s).$$

Let us transform this equation. Since $\psi_m(s)$ is a Taylor polynomial of the mth degree of the function e^s, and for any $m \in \mathbb{N}$ and for all $s \in \mathbb{R}$,

$$\frac{d\psi_m(s)}{ds} = \psi_{m-1}(s), \tag{B.4}$$

the equation is reduced to the form (B.3). Let us study the set of solutions of equation (B.3) on the interval $[0, +\infty)$. To do this, consider a family of $n-1$ functions $\overline{\psi}_k(s) = \theta\psi_k(s) - \psi_{k-1}(s)$, $k = 1, 2, \ldots, n-1$. The functions $\overline{\psi}_k(s)$ have the following property.

Since $\alpha < -\alpha_0$ and, moreover, for any matrix and any type of the norm $2\|H\| > -\alpha_0$, we have $\theta = \frac{\alpha}{2\|H\|} < 1$ for any admissible values of α. And since $\theta < 1$ and $\psi_m(0) = 1$ for any $m \in \mathbb{N}$, the following inequalities hold for the functions $\overline{\psi}_k(s)$:

$$\overline{\psi}_k(0) = \theta - 1 < 0, \quad k = 1, 2, \ldots, n-1. \tag{B.5}$$

On the other hand, from the limit equalities $\lim_{s \to +\infty} \frac{\theta\psi_k(s)}{\psi_{k-1}(s)} = +\infty$, $k = 1, 2, \ldots, n-1$, it follows that there exists a number $\overline{s}_k > 0$ such that for all $s > \overline{s}_k$,

$$\overline{\psi}_k(s) > 0, \quad k = 1, 2, \ldots, n-1. \tag{B.6}$$

Inequalities (B.5) and (B.6) and the continuity of $\overline{\psi}_k(s)$ imply that on the interval $(0, \overline{s}_k)$, the function $\overline{\psi}_k(s)$ has at least one zero. Outside this interval, the function $\overline{\psi}_k(s)$ has no zeros. Denote $\overline{s} = \max\{\overline{s}_1, \overline{s}_2, \ldots, \overline{s}_{n-1}\}$. If for some k from the set $\{1, 2, \ldots, n-2\}$, the function $\overline{\psi}_k(s)$ has a unique zero on the interval $(0, \overline{s})$, then the function $\overline{\psi}_{k+1}(s)$ on $(0, \overline{s})$ also has a single zero. Indeed, let $\overline{\psi}_k(s)$ have a unique zero on $(0, \overline{s})$. Then, since

$$\frac{d\overline{\psi}_{k+1}(s)}{ds} = \theta\frac{d\psi_{k+1}(s)}{ds} - \frac{d\psi_k(s)}{ds} = \theta\psi_k(s) - \psi_{k-1}(s) = \overline{\psi}_k(s),$$

$\overline{\psi}_{k+1}(s)$ has a single extremum point on the interval $(0, \overline{s})$. Moreover, up to this point, $\overline{\psi}_{k+1}(s)$ decreases, and after that, it increases to $+\infty$. Therefore, $\overline{\psi}_{k+1}(s)$ does have a unique zero on $(0, \overline{s})$.

Let us now set $k = 1$. The function $\overline{\psi}_1(s) = \theta\psi_1(s) - \psi_0(s) = \theta(1 + s) - 1 = (\theta - 1) + \theta s$ has a single zero on $(0, \overline{s})$. Then, according to what was proved above, the functions $\overline{\psi}_2(s), \overline{\psi}_3(s), \ldots, \overline{\psi}_{n-2}(s)$ and $\overline{\psi}_{n-1}(s) = \theta\psi_{n-1}(s) - \psi_{n-2}(s)$ also have a single zero on $(0, \overline{s})$, and hence on $[0, +\infty)$. That is, equation (B.3) has a unique solution s^* on the semiaxis $[0, +\infty)$. According to the sufficient extremum conditions, we find that $s = s^*$ is the maximum point of the function $F(s)$ on the interval $[0, +\infty)$. So $\max_{[0,\infty)} F(s) = F(s^*)$, and $\|e^{Ht}\| \leq F(s^*)e^{(\alpha_0 + \alpha)t}$ for all $t \in [0, \infty)$, which was to be proved.

References

[1] Abdeljawad T. (2015). On conformable fractional calculus, *The Journal of Computational and Applied Mathematics*, 279, 57–66.

[2] Alami N. (1986). *Analyse et Commande Optimale des Systèmes Bilinéaires Distribués*. Applications aux Procédés Energétiques. PhD thesis, Univesité de Perpignan, France, Doctorat d'Etat.

[3] Aleksandrov A. Y. (2004). *Stability of Motions of Non-Autonomous Dynamical Systems*, St. Petersburg: Publishing House of St. Petersburg un-ta.

[4] Aleksandrov A. Y. and Platonov A. V. (2012). *Method of Comparison and Stability of Motions of Nonlinear Systems*. SPb: Izd-vo SPb un-ta.

[5] Alefeld G. and Mayer G. (2000). Interval analysis: Theory and applications, *Journal of Computational and Applied Mathematics*, 121, 421–464.

[6] Aminov A. B. (1991). Stability of automatic control systems with a polynomial model, *Automation and Telemechanics*, 10, 44–50.

[7] Anderson D. R. and Georgiev S. (2022). *Conformable Dynamic Equations on Time Scales*, Chapman and Hall/CRC.

[8] Artstein Z. (1983). Stabilization with relaxed controls, *Nonlinear Analysis: Theory, Methods and Applications*, 17(11), 1163–1173.

[9] Babadzhanyants L. K. (2009). Method of additional variables, Bulletin of St. Petersburg University, *Series*, 10(4), 3–11.

[10] Babenko E. A. and Martynyuk A. A. (2016). Stabilization of the motion of a nonlinear system with interval initial conditions, *International Applied Mechanics*, 52(3), 182–191.

[11] Babenko E. A. and Martynyuk A. A. (2016). Stabilization of the motion of affine systems, *International Applied Mechanics*, 52(4), 413–421.

[12] Bainov D. and Simeonov P. (1992). *Integral Inequalities and Applications*, Academic Publishers, Dordrecht.

[13] Banas J. and Goebl K. (1980). *Measures of Noncompactness in Banach Spaces*, New York: Marcel Dekker.

[14] Bellman R. (1943). The stability of solutions of linear differential equations, *Duke Mathematical Journal*, 10, 643–647.

[15] Bellman R. (1953) *Stability Theory of Differential Equations*, New York: Dover Publications.

[16] Benallou A., Mellichamp D. and Seborg D. (1983) Characterisation of equilibrium sets for bilinear systems with feedback control. *Automatica*, 19, 183–189.

[17] Benkhettou N., Hassani S. and Torres D. F. M. (2016). A conformable fractional calculus on arbitrary time scales, *Journal of King Saud University — Science*, 28, 93–98.

[18] Beesack P. (1995). *Gronwall Inequalities*, Carleton Mathematical Lecture Notes, 11.

[19] Bihari I. (1956). A generalization of a lemma of Bellman and its application to uniqueness problems of differential equations, *Acta Mathematica Academiae Scientiarum*, Hungaricae, 7, 71–94.

[20] Bohner M., Peterson A. (2001). *Dynamic Equations on Time Scales: An Introduction with Applications*, Boston: Birkhäuser.

[21] Bohner M. and Martynyuk A. (2024). Equilibrium stability under nuclear confrontation. Differential Equations and Dynamical Systems, 1–15.

[22] Bohner M. and Martynyuk A. (2007). Elements of stability theory of A. M. Liapunov for dynamic equations on time scales, *Nonlinear Dynamics and Systems Theory. An International Journal of Research and Surveys*, 7(3), 225–251.

[23] Bogolyubov N. N. and Zubaryev D. N. (1955). The method of asymptotic approximation for systems with rotating phase and its application to the motion of charged particles in a magnetic field, *Ukrainian Mathematical Journal*, 7, 5–17.

[24] Brauer F. (1963). Bounds for solutions of ordinary differential equations, *Proceedings of the American Mathematical Society*, 14(1), 36–43.

[25] Brogliato B., Lozano R., Maschke B. and Egeland O. (2020). *Dissipative Systems Analysis and Control: Theory and Application*, Berlin: Springer Nature Switzerland AG.

[26] Burton T. A. (1985). *Stability and Periodic Solutions of Ordinary and Functional Differential Equations*, Orlando: Academic Press, Inc.

[27] Burton T. A. (2012). *Liapunov Theory for Integral Equations with Singular Kernels and Fractional Differential Equations*, Port Angeles, WA: Independent Publishing Platform.

[28] Bylov B. F., Vinograd R. E., Grobman D. M. and Nemytsky V. V. (1966). *Lyapunov Exponent Theory*, Moscow: Nauka.

[29] Caputo M. (1969). *Elasticita e Dissipazione*, Bologna: Zanichelli.

[30] Caspary W. R. (1967). Richardson's model of arms races: Description, critique, and an alternative model, *International Studies Quarterly*, 11(1), 63–88.

[31] Coddington E. A. and Levinson N. (1955). *Theory of Ordinary Differential Equations*, New York: McGraw-Hill.

[32] Corduneanu C. (2009). The contribution of R.Conti to the comparison method in differential equations, *Libertas Mathematica*, XXIX, 113–115.

[33] Constanda C., Riva M. D., Lamberti P. D. and Musolino P., eds. (2017). *Integral Methods in Science and Engineering, Theoretical Techniques*, Vol. 1, Berlin, Springer.

[34] Chetaev N. G. (1962). *Movement Stability. Works on Analytical Mechanics*, Moscow: Publishing House of the Academy of Sciences of the USSR.

[35] Chezary L. (1964). *Asymptotic Behavior and Stability of Solutions of Ordinary Differential Equations*, M: Mir.

[36] Cruz-Hernandez C., Martynyuk A. A. and Mazko A. G. (Eds.). (2021). *Advances in Stability and Control Theory for Uncertain Dynamical Systems*, Cambridge: Cambridge Scientific Publishers Ltd.

[37] Echchatbi A., Alami N. and Bouaziz A. (2014). Stabilization and Observation of Bilinear Uncertain Systems, Manuscript, Ecole Mohamadia des Ingenieurs, Morocco.

[38] España M. and Landau I. (1977). Reduced order bilinear models for distillation columns. *Automatica*, 14, 345–355.

[39] Filatov A. N. (1971). *Averaging Methods in Differential and Integro-Differential Equations*, Tashkent: Publishing house "FAN".

[40] Fieguth P. (2021). *An Introduction to Complex Systems: Society, Ecology, and Nonlinear Dynamics*, Berlin: Springer Nature Switzerland AG.

[41] Frangos C. (2021). *Mathematical Modelling, Nonlinear Control and Performance Evaluation of a Ground Based Mobile Air Defence System*, Berlin: Springer Nature Switzerland AG.

[42] Freedman L. and Michaels J. (2019). *The Evolution of Nuclear Strategy*: New, Updated and Completely Revised, Palgrave Macmillan: London.

[43] Golubentsev A. N. (1967). *Integral Methods in Dynamics*, Kiev, Tekhnika.

[44] Gleditsch N. P., ed. (2021). *Lewis Fry Richardson: His Intellectual Legacy and Influence in the Social Sciences*, vol. 27 of Pioneers in Arts, Humanities, Science, Engineering, Practice. Springer, Cham.

[45] Gutowski R. and Radziszewski B. (1970). Asymptotic behaviour and properties of solutions of a system of nonlinear second order ordinary differential equations describing motion of mechanical systems. *Archiwum Mechaniki Stosowanej*, 6(22), 675–694.

[46] Grebenikov E. A. and Riabov Yu. A. (1971). *New Qualitative Methods in Celestial Mechanics*, Moscow: Nauka.

[47] Gronwall T. H. (1919). Note on the derivatives with respect to a parameter of the solutions of a systems of a system of differential equation, *Annals of Mathematics*, 20(2), 292–296.

[48] Hahn W. (1967). *Stability of Motion*, Berlin: Springer-Verlag.

[49] Heuser B. (2024). Return to a bleak past? The Russian Invasion of Ukraine and the International System, in Janne Haaland Matlary and Rob Johnson (eds.): *NATO and the Russian War in Ukraine: Strategic Integration and Military Interoperability* (London: Hurst, 99–116).

[50] Hilger S. (1990). Analysis on measure chains-a unified approach to continuous and discrete calculus, *Results in Mathematics*, 18, 18–56.

[51] Hale J. (1984). *Theory of Functional-Differential Equations*, Moscow: Mir.

[52] Hart W. Z. (1917). Differential equations and implicit function with infinite, *Transactions of the American Mathematical Society*, 18, 125–160.

[53] Hayashi C. (1966). *Nonlinear Oscillations in Physical Systems*, New York: McGraw-Hill.

[54] Hubbard J. H. and West B. H. (1995). *Differential Equations: A Dynamical Systems Approach: Higher-Dimensional Systems* (Texts in Applied Mathematics, 18), Berlin: Springer.

[55] Kamenkov G. V. (1972). *Stability and Oscillations of Nonlinear Systems*, Moscow: Nauka.

[56] Kim A. V. (1992). *Lyapunov's Direct Method in the Theory of Stability of Systems with Aftereffect*, Yekaterinburg: Publishing House Ural State University.

[57] Kardous Z. and Benhadj Braiek N. (2014). Stabilizing Sliding Mode Control for Homogeneous Bilinear Systems, *Nonlinear Dynamics and Systems Theory*, 14(3), 303–312.

[58] Keller S. (2000). *Asymptotisches Verhalten invarianter Faserbündel bei Diskretisierung und Mittelwertbildung im Rahmen der Analysis auf Zeitskalen*, Ph.D. thesis, Universität Augsburg.

[59] Kilbas A., Srivastava M. H. and Trujillo J. J. (2006). *Theory and Applications of Fractional Differential Equations*, Amsterdam: North Holland.

[60] Khalil R., Horani M. A., Yousaf A. and Sababhen M. (2014). A new definition of fractional derivative, *Journal of Computational and Applied Mathematics*, 264, 65–70.

[61] Krasovsky A. A. (1973). *Automatic Flight Control Systems and Their Analytical Design*, Moscow: Nauka, 558 p.

[62] Krasovsky N. N. and Subbotin A. I. (1974). *Positional Differential Games*, Moscow: Nauka.

[63] Krylov A. N. and Bogoliubov N. N. (1937). *Introduction to Nonlinear Mechanics*, Kyiv: Publishing House of the Academy of Sciences of the Ukrainian SSR.

[64] Kuratowski K. (1966). *Topology*, Vol. 2, New York: Academic Press.

[65] Lakshmikantham V., Leela S. and Devi J. V. (2009). *Theory of Fractional Dynamic Systems*. Cambridge: Cambridge Scientific Publishers.

[66] Lakshmikantham V., Leela S. and Martynyuk A. (2015). *Stability Analysis of Nonlinear Systems*, Second Edition. Berlin: Birkhäuser.

[67] Lakshmikantham V., Leela S. and Martynyuk A. A. (1990). *Practical Stability of Nonlinear Systems*, Singapore: World Scientific, 207 p.

[68] Lakshmikantham V., Leela S. and Vatsala A. (2011). *Theory of Differential Equations in Cones*. Cambridge: Cambridge Scientific Publishers.

[69] Langenhop C. E. (1960). Bounds on the norm of a solution of a general differential equation, *Proceedings of the American Mathematical Society*, 11, 795–799.

[70] La Salle J. and Lefschetz S. (1961). *Stability by Liapunov's Direct Method with Applications*, New York: Academic Press.

[71] Longchamp R. (1980). Stable feedback control of bilinear systems. *IEEE Transactions on Automatic Control*, 25, 302–306.

[72] Louartassi Y., El Mazoudi E. H. and El Alami N. (2012). A new generalization of lemma Gronwall-Bellman, *Applied Mathematical Sciences*, 6(13), 621–628.

[73] Lee E. B. and Markus L. (1972). *Fundamentals of the Theory of Optimal Control*, Moscow: Nauka.

[74] Leitmann G. (1966). *An Introduction to Optimal Control*, New York: McGraw-Hill.

[75] Lungu N. and Ciplea S. A. (2017). Optimal Gronwall Lemmas, *Fixed Point Theory*, 18(1), 293–304.

[76] Lyapunov A. M. (1950). *The General Problem of Motion Stability*, M.-L.: GITTL.

[77] Malkin I. G. (1938). On the stability of motion in the sense of Lyapunov, *Matematicheskii Sbornik*, 3(45), 47–100.

[78] Martynyuk A. A. (1994). On a generalization of Richardson's model of the arms race. Rossiĭskaya Akademiya Nauk, *Doklady Akademii Nauk*, 339(1), 15–17.

[79] Martynyuk A. A. (2016). Stability theory for dynamic equations on time scales. *Systems & Control: Foundations & Applications*, Birkhäuser/Springer [Cham].

[80] Martynyuk A., Lakshmikantham V. and Leela S. (1989). *Stability of Motion: Method of Integral Inequalities*, Kyiv: Naukova Dumka.

[81] Martynyuk A. A. (1973). *Technical Stability in Dynamics*, Kyiv: Technique.

[82] Martynyuk A. A. (2015). On a method for estimating solutions of quasilinear systems, *Dopovidi Natsional'noi Akademii Nauk Ukrayiny*, 2, 19–23.

[83] Martynyuk A. A. (2015). Novel Bounds for Solutions of Nonlinear Differential Equations, *Applied Mathematics*, 6, 182–194.

[84] Martynyuk A. A. (1995). *Stability Analysis: Nonlinear Mechanics Equations*, Amsterdam: Gordon and Breach Publishers.

[85] Martynyuk A. A. (1970). To one method of studying mechanical systems with distributed parameters, *Prikladnaya Mekhanika*, 6(12), 97–103.

[86] Martynyuk A. A. (2007). *Stability of Motion: The Role of Multicomponent Lyapunov's Functions*, Cambridge: Cambridge Scientific Publishers.

[87] Martynyuk A. A. (2018). On the stability of fractional-like systems of equations of perturbed motion, *Dopovidi Natsional'noi Akademii Nauk Ukrayiny*, 6, 9–16.

[88] Martynyuk A. A. and Babenko E. A. (2016). Finite time stability of uncertain affine systems, *Mathematics in Engineering, Science, and Aerospace*, 7(1), 179–196.

[89] Martynyuk A. A. and Babenko E. A. (2017). Robust stabilization of bilinear systems under interval initial conditions, *International Applied Mechanics*, 53(4), 454–463.

[90] Martynyuk A. A. (2016). *Stability Theory for Dynamic Equations on Time Scales*. Berlin: Birkhäuser.

[91] Martynyuk A. A. (2023). Analysis of equi-boundedness and stability of essentially nonlinear systems, *International Applied Mechanics*, 59(1), 69–78.

[92] Martynyuk A. A. (2023). Stability and boundedness of solutions of dynamic equations with conformable derivative of the state vector, *International Applied Mechanics*, 59(6), 631–640.

[93] Martynyuk A. A. and Martynyuk-Chernienko Y. A. (2020). Boundedness of solutions of conformable fractional equations of perturbed motion, *International Applied Mechanics*, 56(5), 572–580.

[94] Martynyuk A. A., Stamova I. and Stamov, G. T. (2018). On a development of the comparison principle in the stability theory of motion, *Libertas Mathematica*, 38(2), 41–55.

[95] Martynyuk, A. A. and Stamova I. M. (2018). Fractional-like derivative of Lyapunov-type functions and applications to stability analysis of motion, *Electronic Journal of Differential Equations*, 62, 1–12.

[96] Martynyuk, A. A., Stamov, G. and Stamova, I. M. (2019). Practical stability analysis with respect to manifolds and boundedness of differential equations with fractional-like derivatives, *Rocky Mountain Journal of Mathematics*, 49(1), 211–233.

[97] Martynyuk A. A., Stamova I., Stamov G. T. and Martynyuk-Chernienko, Y. A. (2022). On the boundedness and Lagrange stability of fractional-like neural network-based quasilinear systems, *The European Physical Journal Special Topics*, 231, 1789–1799.

[98] Martynyuk A. A., Stamov, G. I. and Stamova, I. M. (2019). Impulsive fractional-like differential equations: practical stability and boundedness with respect to h-manifolds, *Fractal and Fractional (MDPI)*, 3(4), 50.

[99] Martynyuk A., Stamov G., Stamova I. and Gospodinova E. (2023). Formulation of Impulsive Ecological Systems using the Conformable Calculus Approach: Qualitative Analysis (MDPI), *Mathematics*, 11(10), 1–16.

[100] Martynyuk, A. A., Radziszewski B. and Szadkowski, A. (2020). *Stability: Elements of the Theory and Applications with Examples*, Warsaw: De Gruyter/SCIENDO.

[101] Martynyuk A. A., Chernetskaya L. N. and Martynyuk V. A. (2013). *Weakly Connected Nonlinear Systems. Boundedness and Stability of Motion*. Boca Raton: CRC Press Taylor & Francis Group.

[102] Martynyuk A. A. and Chernienko V. A. (2020). Sufficient conditions for the stability of motion of polynomial systems, *International Applied Mechanics*, 56(1), 13–21.

[103] Martynyuk A. A. and Chernienko V. A. (2021). On estimate of Lyapunov function and stability of motion of system with asymptotic expansion of right-hand side of equations of perturbed motion, *International Applied Mechanics*, 57(1), 11–18.

[104] Martynyuk A. A. and Chernienko V. A. (2021). On stabilization of motion of non-autonomous polynomial system, *International Applied Mechanics*, 57(5), 524–533.

[105] Martynyuk A. A. and Chernienko V. A. (2022). On the stability of the motion of polynomial systems with aftereffect, *International Applied Mechanics*, 58(4), 373–380.

[106] Martynyuk A. A. and Chernienko V. A. (2021). To the problem of the stability of the movement of purely non-linear systems, *Dopovidi Natsional'noi Akademii Nauk Ukrayiny*. 2, 3–12.

[107] Martynyuk A. A., Khusainov D. Y. and Chernienko V. A. (2016). Integral estimates of solutions to nonlinear systems and their applications, *Nonlinear Dynamics and Systems Theory*, 16(1), 1–11.

[108] Martynyuk A. A., Khusainov D. Y. and Chernienko V. A. (2017). Constructive estimation of the lyapunov function for quadratic nonlinear systems, *International Applied Mechanics*, 53(3), 346–357.

[109] Mazko A. G. (2022). Weighted damping of external and initial disturbances in descriptor control systems, *Ukrainian Mathematical Journal*, 73(10), 1590–1606.

[110] Mazko A. G. (2022). Weighted performance measure and generalized H_∞ control problem for linear descriptor systems, *Nonlinear Dynamics and Systems Theory*, 22(3), 303–318.

[111] Mazko A. G. (2023). *Matrix Methods for the Analysis and Synthesis of Dynamical Systems*, Kyiv: Naukova dumka.

[112] Mazko A. G. (2025). Evaluation of the weighted level of attenuation of external and initial disturbances in nonlinear systems, *Ukrainian Mathematical Journal*, 76(8), 1338–1351.

[113] Metz D. and Viorel A. (2020). Nonlinear economic growth dynamics in the context of a military arms race, *Studia Universitatis Babeş-Bolyai, Mathematica*, 65(2), 309–320.

[114] Peterson A. C. and Tisdell C. C. Boundedness and uniqueness of solutions to dynamic equations on time scales, *Journal of Difference Equations and Applications*, 10(13–15), 1295–1306.

[115] Potzsche C. (2002). Chain rule and invariance principle on measure chains, dynamic equations on time scales, *Journal of Computational and Applied Mathematics*, 141, 249–254.

[116] Miller R. K. (1971). *Nonlinear Volterra Integral Equations*, Menlo Park, W. A. Benjamin, Inc.

[117] Mishina A. P. and Proskuryakov I. V. (1965). *Highest Algebra*, Moscow: Nauka.

[118] Mursaleen M. and Mohiuddine S. A. (2012). Applications of measures of noncompactness to the infinite system of differential equations in l_p spaces, *Nonlinear Analysis*, 75, 2122–2115.

[119] Melnikov G. I. (1956). *Some Questions of Lyapunov's Direct Method*, Dokl. USSR Academy of Sciences, 110(3), 326–329.

[120] Moiseev N. D. (1949). *Essays on the Development of the Theory of Stability.* M.-L.: Gostekhizdat.

[121] Mohler R. (1991). *Nonlinear Systems: Applications to Bilinear Control*, vol. 2. Englewood Cliffs, New Jersey: Prentice Hall.

[122] Mohler R. (1973). *Bilinear Control Processes: With Applications to Engineering, Ecology, and Medicine*, New York and London: Academic Press.

[123] Moore R. E. (1966). Interval Analysis, Prentice-Hall, Englewood Cliffs, NJ.

[124] N'Doye I. (2011). *Generalization du lemme de Gronwall-Bellman pour la stabilisation des systèmes fractionnaires*, These, Nancy-Universite.

[125] N'Doye I., Zasadzinski M., Darouach M., Radhy N. E. and Bouaziz A. (2011). Exponential stabilization of a class of nonlinear systems: A generalized Bellman-Gronwall lemma approach, *Nonlinear Analysis: Theory, Methods and Applications*, 74(18), 7333–7341.

[126] Newman P. K. (1959). Some notes on stability conditions, *The Review of Economic Studies*, 27(1), 1–9.

[127] Nosov V. R., Ortega Herrera J. A. and Dominiguez H. (2009). Stability of some polynomial equations with delay, *Functional Differential Equations*, 16(3), 561–578.

[128] Oguntuase J. A. (2001). On an Inequality of Gronwall, *Journal of Inequalities in Pure and Applied Mathematics*, 2(1), 1443–5756.

[129] Pachpatte B. G. (1998). *Inequalities for Differential and Integral Equations*, San Diego, ets.: Academic Press.

[130] Pal D. and Mahapatra G. S. (2016). Dynamic behavior of a predator-prey system of combined harvesting with interval-valued rate parameters, *Nonlinear Dynamics*, 83(4), 2113–2123.

[131] Pal D., Mahapatra G. S. and Samanta G. (2016). Stability and bionomic analysis of fuzzy prey-predator harvesting model in presence of toxicity: A dynamic approach, *Bulletin of Mathematical Biology*, 78(7), 1493–1519.

[132] Pal D., Mahapatra G. S. and Samanta G. (2013). Optimal harvesting of prey-predator system with interval biological parameters: A bioeconomic model. Mathematical Biosciences, 241(2), 181–187.

[133] Pandey S. C. and Raturi A. K. (2023). On solutions to the arms race model using some techniques of fractional calculus. *Journal of the Ramanujan Society of Mathematics and Mathematical Sciences*, 10(2), 45–60.

[134] Patan K. (2019). *Robust and Fault-Tolerant Control: Neural-Network-Based Solutions*, Berlin: Springer Nature Switzerland AG.

[135] Peano G. (1885–1886). Sull' integrabilitá delle equazioni differenziali del primo ordine, *Atti della Reale Accademia delle Scienze di Torino*, 21, 677–685.

[136] Persidskii K. P. (1976). *Infinite Systems of Differential Equations. Differential Equations in Nonlinear Spaces*, Alma-Ata: Nauka.

[137] Poincaré A. (1971). *Selected Works*, Volume 1, Moscow: Nauka.

[138] Poincaré A. (1947). *On Curves Defined by Differential Equations*, Moscow: Gostekhizdat.

[139] Podlybny I. (1999). *Fractional Differential Equations*, London, Academic Press.

[140] Roseau M. (1971). *Nonlinear Oscillations and Stability Theory*. Moscow: Nauka.

[141] Rahmat M. R. S. (2019). A new definition of conformable fractional derivative on arbitrary time scales, *Advances in Differential Equations* (2019), 1–16.

[142] Rama Mohana Rao M. (1980). *Ordinary Differential Equations*, New Delhi-Madras: Affiliated East-West Press Pvt Ltd.

[143] Raffoul Y. (2003). Boundedness in nonlinear differential equations, *Nonlinear Studies*, 10(4), 1–9.

[144] Reid W. T. (1930). Properties of solutions of an infinite systems of ordinary linear differential equations of the first order with auxiliary boundary conditions, *Transactions of the American Mathematical Society*, 1930, 284–318.

[145] Richardson L. F. (1960). *Arms and Insecurity: A Mathematical Study of the Causes and Origins of War*. The Boxwood Press, Pittsburgh, Pa.; Quadrangle Books, Chicago, Ill. Edited by Nicolas Rashevsky and Ernesto Trucco.

[146] Roxin E. O. (1977). *Control Theory and its Applications*, Amsterdam: Gordon and Breach Science Publishers.

[147] Rogalev A. N. (2004). Boundaries of solution sets for systems of ordinary differential equations with interval initial data, *Computational Technologies*, 9(1), 86–94.

[148] Rumiantsev V. V. and Oziraner A. S. (1987). *Stability and Stabilization of Movement in Relation to Some of the Variables*, Moscow: Nauka.

[149] Ryan E. and Buckingham N. (1983). On asymptotically stabilizing feedback control of bilinear systems, *IEEE Transactions on Automatic Control*, 28, 863–864.

[150] Samoilenko A. M. and Teplinskii Y. V. (2003). *Countable Systems of Differential Equations*, Utrecht-Boston: VSP.

[151] Santra P. K. and Mahapatra G. S. (2020). Dynamical study of discrete-time prey-predator model with constant prey refuge under imprecise biological parameters, *Journal of Biological Systems*, 28(3), 681–699.

[152] Šiljak D. D. (1976). Competitive analysis of the arms race, In *Annals of Economic and Social Measurement*, vol. 5 of NBER Chapters, 283–295.

[153] Šiljak D. D. (1977). On the stability of the arms race. In John V. Gillespie and Dina A. Zinnes, editors, *Mathematical Systems in International Relations Research*, Chapter 9, 264–304. Praeger Publishers, New York.

[154] Šiljak D. D. (1979). Large-Scale Dynamic Systems, Stability and Structure, vol. 3 of North-Holland Series in System Science and Engineering. North-Holland Publishing Co., New York-Amsterdam.

[155] Sirazetdinov T. K. and Aminov A. B. (1984). On the problem of constructing Lyapunov functions in the study of global stability of solutions to systems with a polynomial right-hand side, In *The Method of Lyapunov Functions and its Applications* (Eds. V. M. Matrosov, S. N. Vasiliev), Novosibirsk: Nauka, 72–74.

[156] Shary S. P. (2020). *Finite-Dimensional Interval Analysis*, publishing house "XYZ" in Novosibirsk.

[157] Schumacher J. M. (1983). A direct approach to compensator design for distributed parameter systems, *SIAM Journal on Control and Optimization*. 21, 823–836.

[158] Skullestad A. and Gilbert J. M. (2000). H_∞ control of a gravity gradient stabilised satellite, *Control Engineering Practice*, 8(9), 975–983.

[159] Slemrod M. (1978). Stabilization of bilinear control systems with applications to nonconservative problems in elasticity, *SIAM Journal on Control and Optimization*, 16, 131–141.

[160] Slyusarchuk V. Y. (2003). *Absolute Stability of Dynamic Systems with Aftereffect.* Rivne: Ukrainian holding "University of Waters and Natural Resources".

[161] Smith R. P. (2021). The influence of the Richardson arms race model. In *Lewis Fry Richardson: His Intellectual Legacy and Influence in the Social Sciences*, volume 27 of Pioneers Arts Humanit. Sci. Eng. Pract., pages 25–34. Springer, Cham.

[162] Sontag E. (1999). Stability and stabilization: Discontinuities and the effect of disturbances, In *Nonlinear analysis, Differential Equations, and Control* (F. H. Clarke and R. J. Stern, eds.), 551–598.

[163] Sun Y. and Guo L. (2006). *On Controllability of Some Classes of Affine Nonlinear Systems*, Sun Yat-Sen University (manuscript), 1–21.

[164] Tikhonov A. N. (1934). Über Unendliche Systeme von Differentialgleichungen, *Mathematical collection*, 41(1), 551–560.

[165] Tonkov E. L. (1972). *Stability of Solutions to Ordinary Differential Equations.* Moscow: Publishing House of Moscow Institute of Chemical and Engineering.

[166] Veretennikov V. G. (1984). Stability and Oscillations of Nonlinear Systems, Moscow: Nauka.

[167] Van der Waerden B. L. (1949). *Moderne Algebra*, Berlin: Verlag von Julius Springer.

[168] Younus A., Bukhsh K., Alqudah M. A. and Abdeljawad T. (2022). Generalized exponential function and initial value problem for conformable dynamic equations, *AIMS Mathematics*, 7(7), 12050–12076.

[169] Yoshizawa T. (1966). *Stability Theory by Lyapunov's Second Method*, Tokyo: The Mathematical Society of Japan.

[170] Wang Y. N., Zhou J. W. and Li Y. K. (2016). Fractional Sobolev's spaces on time scales via conformable fractional calculus and their application to a fractional differential equation on time scales, *Advances in Mathematical Physics* (2016), 1–21.

[171] Yuan A., Joubert S. V. and Gai Y. (2008). Applications of ODEs to the mathematical modelling of international conflict. In Buffelspoort TIME peer-reviewed conference proceedings, 22–26 September, 187–197.

[172] Zecevic A. I. and Siljak D. D. (2003). Stabilization of nonlinear systems with moving equilibria, *IEEE Transactions on Automatic Control*, 48(6) 1036–1040.

[173] Zheng A., Feng Yu. and Wang W. (2015). The Hyers-Ulam stability of the conformable fractional differential equation, *Mathematica Aeterna*, 5(3), 485–492.

[174] Zhautykov O. A. (1965). Averaging principle in nonlinear mechanics as applied to countable systems of equations, *Ukrainskii Matematicheskii Zhurnal*, 17(1), 1965, 39–46.

[175] Zou A.-M. (2014). Finite-time output feedback attitude tracking control for rigid spacecraft, *IEEE Transactions on Control Systems Technology*, 22(1), 338–345.

[176] Zubov V. I. (1959). *Mathematical Methods for Studying Automatic Control Systems*, Leningrad: Sudpromgiz.

[177] Zubov V. I. (2009). *Lectures on Control Theory*, Textbook. SPb, Publishing House "Lan".

Index

Series on Advances in Mathematics for Applied Sciences

Editorial Board

Series on Advances in Mathematics for Applied Sciences

Aims and Scope

This series reports on new developments in mathematical research related to new analytical and numerical methods, mathematical modeling in the applied physical and natural sciences as well as in economics, and quantitative and qualitative analysis approaches for the mathematical models. Topics covered include modelling, constitutive theories, fluid and solid mechanics, kinetic and transport theories. The series ranges from monographs to lecture notes, quality conference proceedings and collections of papers. The high quality of research, the novelty of mathematical tools, and the potential for frontier problems will be the guidelines for the selection of the content for this series.

Instructions for Authors

Submission of proposals should be addressed to the editors-in-charge or to any member of the editorial board. In the latter, the authors should also notify the proposal to one of the editors-in-charge. Acceptance of books and lecture notes will generally be based on the description of the general content and scope of the book or lecture notes as well as on sample of the parts judged to be more significantly by the authors.

Acceptance of proceedings will be based on relevance of the topics and of the lecturers contributing to the volume.

Acceptance of monograph collections will be based on relevance of the subject and of the authors contributing to the volume.

Authors are urged, in order to avoid re-typing, not to begin the final preparation of the text until they received the publisher's guidelines. They will receive from World Scientific the instructions for preparing camera-ready manuscript.

Series on Advances in Mathematics for Applied Sciences

ISSN: 1793-0901

*To view the complete list of the published volumes in the series, please visit:
https://www.worldscientific.com/series/samas